Light and Living Matter: A Guide to the Study of Photobiology

Volume 2: The Biological Part

Roderick K. Clayton, Ph.D.

Professor of Biology and Biophysics
Cornell University

McGraw-Hill Book Company

New York
St. Louis
San Francisco
Düsseldorf
Johannesburg
Kuala Lumpur
London
Mexico
Montreal
New Delhi
Panama
Rio de Janeiro
Singapore
Sydney
Toronto

Light and Living Matter,
Volume 2: The Biological Part

Library of Congress Catalog Card Number 75-114444

1234567890 VBVB 7987654321

This book was set in Times Roman by Vail-Ballou Press, Inc., and printed and bound by Vail-Ballou Press, Inc. The designer was Marsha Cohen; the drawings were done by John Cordes, J. & R. Technical Services, Inc. The editors were Jeremy Robinson and Andrea Stryker-Rodda. Peter D. Guilmette supervised production.

Contents

SERIES PREFACE ix

PREFACE xi

1. PHOTOSYNTHESIS 1

Biochemical Patterns in Photosynthesis 2

1-1. Chemical Oxidation and Reduction 2
1-2. Photosynthesis as a Photochemical Oxidation-reduction Process 5
1-3. Adenosine Triphosphate 10
1-4. Patterns of Electron Transport in Photosynthetic Bacteria and Green Plants 16

Photosynthetic Units: Light-harvesting Systems and Reaction Centers 22

1-5. Photochemical Reaction Centers 22
1-6. The Light-harvesting Antenna 28
1-7. The Diversification of Light-harvesting Systems 31

The Cooperation of Two Photochemical Systems in Photosynthesis 36

1-8. Evidence for the Existence and Cooperation of Two Systems 36
1-9. The Series Formulation, or *z* Scheme 39
1-10. System II: Oxygen Evolution, Electron Transport, and Chlorophyll Fluorescence 43
1-11. The Relationship between Photosynthetic Bacteria and Green Plants 51
1-12. The Dynamics of Photochemistry in the Bacteria 53

Structure and Function: Membranes and Regulatory Systems 54

1-13. The Structures of Photosynthetic Tissues 54

1-14. The Coupling between Electron Transport and Phosphorylation 57
1-15. Uncouplers: Ion Transport and Membrane Potential 61
1-16. Two Hypothetical Regulating Devices 63

Bibliography 64

2. *PHOTOTAXIS AND PHOTOTROPISM: MOVEMENT AND GROWTH IN RELATION TO LIGHT* 67

Phototaxis 68

2-1. Varieties of Phototactic Response 68
2-2. Taxis in Relation to Metabolism 71
2-3. T. W. Engelmann: An Appreciation 75

Phototropism 76

2-4. General Characteristics of Phototropic Systems 76
2-5. The Problem of the Photoreceptor 80

Mechanism 82

2-6. Speculations about Mechanism 82

Bibliography 83

3. *VISION IN HIGHER ANIMALS* 85

Anatomy and Function in the Eye 88

3-1. Anatomy 88
3-2. Rods and Cones 93

The Limits of Visual Acuity and Sensitivity 97

3-3. Visual Acuity 97
3-4. Absolute Sensitivity and Dark Adaptation 100

The Visual Pigments: Their Chemistry and Function 108

3-5. The Visual Pigments 108
3-6. The Chemistry of Vision 113
3-7. The Fast Photovoltage 118
3-8. Color Vision 122

From Quanta to Nerve Impulses: The Transduction and Processing of Visual Information **127**

3-9. Transduction 127
3-10. The Processing of Visual Information in the Retina 131

Bibliography **137**

4. *LIGHT AS IT REGULATES DAILY AND SEASONAL RHYTHMS* **139**

Adapting to Day and Night and to the Seasons **140**

4-1. Daily Rhythms 140
4-2. Seasonal Cycles 141

Some Patterns and Examples of Photoregulation **146**

4-3. Circadian Rhythms in Animals 146
4-4. Seasonal Rhythms in Animals 147
4-5. Mechanisms in the Photoregulation of Animals 148
4-6. General Features of Photoregulation in Plants 150

Phytochrome **156**

4-7. The Chemistry of Phytochrome 156
4-8. The Photophysiology of Phytochrome 162

Rhythms in Plants **165**

4-9. The Nature of the Circadian Clock 165
4-10. The Clock and the Estimation of Day Length by Plants 170

Bibliography **173**

5. *DAMAGE AND THE REPAIR OF DAMAGE BY LIGHT* **175**

The Photochemistry and Photobiology of Nucleic Acids and Proteins 178

5-1. Structure and Photochemistry of DNA and RNA 179
5-2. Photobiology of Nucleic Acids and Proteins 183

Recovery from Damage by Ultraviolet 189

5-3. Primitive Earth Conditions 189
5-4. Analysis of Relations between Dose and Effect in Terms of Target Theory 190
5-5. Dark Repair of DNA 194
5-6. Photoreactivation: Repair of DNA and RNA 196

Damaging Effects of Long-wave Ultraviolet and Visible Light 198

5-7. A Variety of Biological Effects 198
5-8. The Photophysiology of Human Skin 202
5-9. Photochemical Electron Transfer and the Participation of Oxygen 203
5-10. Photochemical Oxygenation: The Nature and Importance of Excited Oxygen 206

Bibliography 213

6. THE LUMINESCENCE OF FIREFLIES AND OTHER LIVING THINGS 215

Diverse Chemical Patterns of Bioluminescence 216

6-1. Luminous Crustaceans, Protozoa, Fungi, and Sea Pansies 216
6-2. Luminous Bacteria 219
6-3. Bioluminescence without Oxygen: Acorn Worms and Jellyfish 221
6-4. Firefly Luminescence 222

The Biological Significance of Luminescence 229

Bibliography 230

INDEX 233

The Chemistry-Biology Interface Series

Several years ago, a few dozen biologists, chemists, physicists, and other scientists spent several days on the campus of the University of Washington under the joint sponsorship of the Commission on Undergraduate Education in Biology, the Advisory Council on College Chemistry, and the Commission on College Physics. The purpose was to study ways to improve teaching in areas of mutual concern to two or more of the disciplines involved. The group considering the area between chemistry and biology agreed that a series of paperback books, prepared for elementary college level students in either biology or chemistry could serve a useful purpose toward this end.

Prepared by authorities in their fields, these books could, for the chemists, indicate the biologically significant reactions useful to illustrate chemical principles and, for the biologist, summarize up-to-date information on molecular phenomena of significance to a modern understanding of biological systems.

To implement this proposal, CUEBS and AC3 appointed an editorial committee of:

Professor Robert H. Burris, Department of Biochemistry, University of Wisconsin
Professor L. Carroll King, Department of Chemistry, Northwestern University
Professor Leonard K. Nash, Department of Chemistry, Harvard University
Professor Aubrey W. Naylor, Department of Botany, Duke University
Professor Charles C. Price, Department of Chemistry, University of Pennsylvania

to organize the undertaking.

As of this writing, the following volumes have been published:

O. T. Benfey, "Introduction to Organic Reaction Mechanisms"
Roderick K. Clayton, "Light and Living Matter," Vols. 1 and 2
Charles C. Price, "Geometry of Molecules"

The following volumes are planned:

Myron Bender, "Catalysis"
Melvin Calvin, "Chemical Evolution"
Paul M. Doty, "Macromolecules"
David E. Greene, "Surface, Films and Membranes"

It is our hope that the material in these volumes will prove of sufficient interest to teachers and students in elementary college chemistry and biology courses that much of it will ultimately be incorporated in regular textbooks.

Charles C. Price
Philadelphia, Pennsylvania

Preface

In conjunction with Volume 1, this book provides for undergraduate students an introduction to some of the methods, principles, facts (as currently believed) and speculations of photobiology. As far as possible the material has been kept at a self-explanatory level, so as to be comprehensible to students whose background is limited to introductory college courses in chemistry, physics and mathematics.

In Volume 2 I have tried to communicate the viewpoint of the investigator by exposing areas of ignorance and by delineating those aspects of photobiology in which research seems especially fruitful at present. At the same time I have tried to present the most salient facts and to introduce the reader to a variety of analytical approaches to the subject. One purpose of this treatment has been to provide a concise framework for a course in photobiology, emphasizing the multidisciplinary character of the subject.

Roderick K. Clayton

1. Photosynthesis

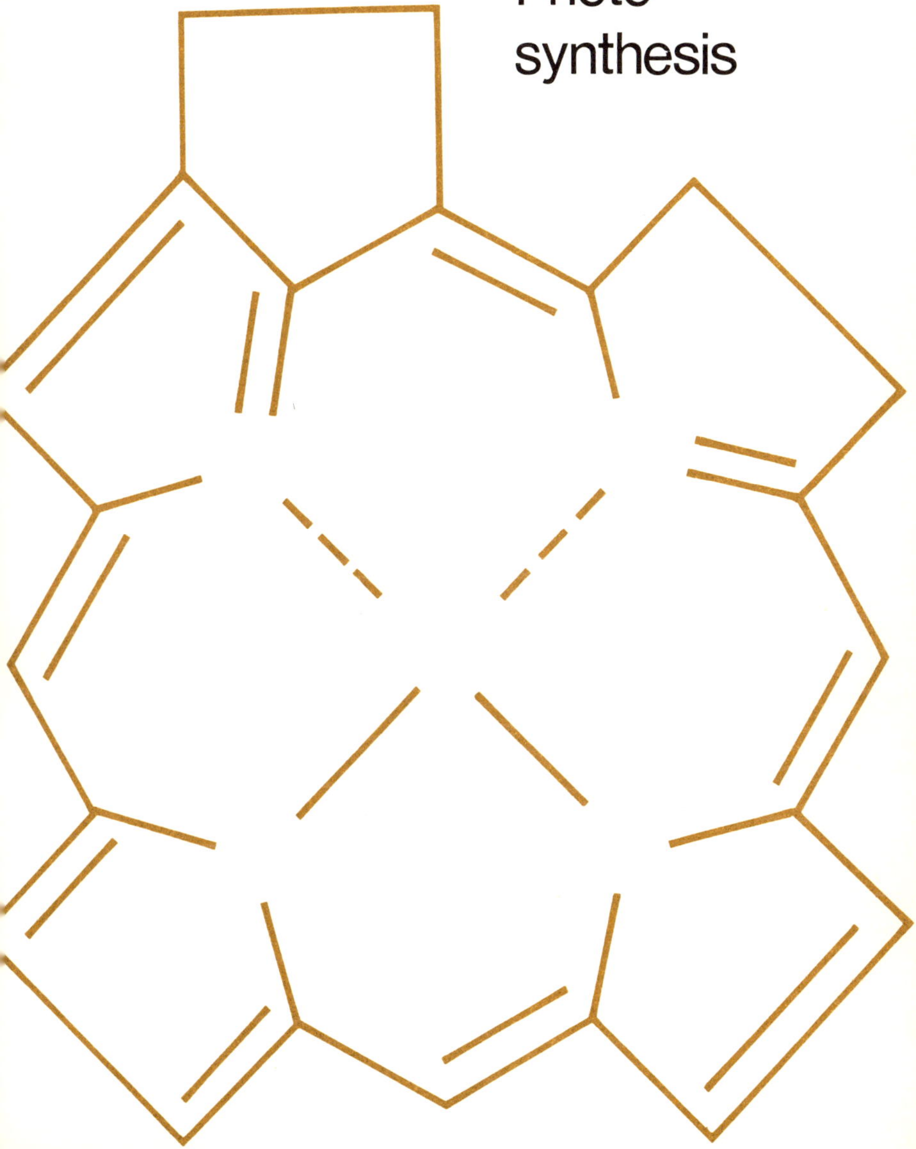

We shall begin the biological part of this series with a consideration of photosynthesis, the process by which plants store the energy of sunlight. The study of photosynthesis will also serve to introduce physical mechanisms and chemical patterns that are important in other aspects of photobiology, to be treated in subsequent chapters.

The essential sequence of events in photosynthesis, to be elaborated in this chapter, is as follows: Chlorophyll (henceforth abbreviated Chl) absorbs light and sensitizes a photochemical oxidation-reduction reaction of the kind shown in Fig. 1-1 (this is the same as Fig. 2-26 in Volume 1). The reducing side of the process ultimately brings about the conversion of carbon dioxide to sugar. The oxidizing side produces oxygen from water in green plant photosynthesis. In a variation of this pattern the oxidant and reductant interact, and the energy of the interaction becomes stored chemically and later is used for a variety of purposes. The main energy-storing substance is adenosine triphosphate (ATP).

A distinctive mechanism in photosynthesis is found in the light-harvesting stage of the process. Most of the Chl (and other pigment) in a plant is not involved directly in the photochemistry; rather it forms an antenna for absorbing light quanta. The quanta are transferred, probably in the manner described in Sec. 2-17 in Volume 1, to a small fraction of Chl molecules that are specialized so as to participate in the photochemical reaction.

BIOCHEMICAL PATTERNS IN PHOTOSYNTHESIS

1-1. Chemical Oxidation and Reduction

Because photosynthesis begins as a light-driven oxidation-reduction process, it may be helpful to start with some simple remarks about oxidation and reduction. These terms refer to the loss or gain of electrons by an atom or molecule, in a reaction like

$$A + D \longrightarrow A^- + D^+ \qquad (1\text{-}1)$$

where A becomes reduced and D becomes oxidized. This reaction involves a separation of negative (—) and positive (+) charges

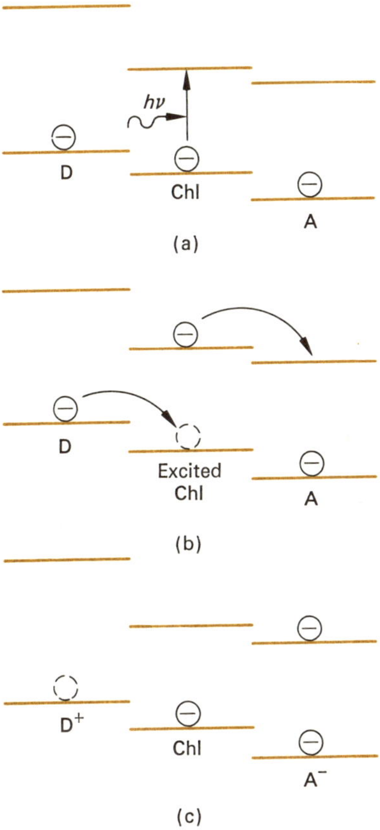

Fig. 1-1. Chlorophyll (Chl) can mediate the photochemical transfer of an electron from a donor molecule D to an acceptor A, forming oxidized donor D^+ and reduced acceptor A^-. See Sec. 2–13 in Volume 1.

against their mutual attraction. It therefore requires an input of energy which becomes stored in the products. This is analogous to the energy needed to raise a stone against the earth's gravitational attraction. Then the reverse process

$$A^- + D^+ \longrightarrow A + D$$

liberates an equal amount of energy as negative and positive charges are allowed to come together. In the gravitational analogy a stone falls, and the gravitational potential energy is released as kinetic energy. Now the energy may be dissipated, as when the stone strikes the ground and heats it, or it may be conserved. The stone might fall on a spring-loaded platform with a latch, so that energy could be stored in the compressed spring. Similarly the reaction $A^- + D^+ \longrightarrow A + D$ might be coupled to another chemical reaction that requires energy. Then the energy released in the first reaction could be stored as chemical bond energy in products of the second.

The combustion of sugar with oxygen is an example of an energy-yielding oxidation-reduction process, in which oxygen is reduced and sugar is oxidized. The reverse process of green plant photosynthesis, in which carbon dioxide and water are converted to sugar and oxygen, might then be regarded overall as an energy-requiring oxidation-reduction, in which light provides the energy. A simpler kind of light-driven and energy-storing oxidation-reduction reaction is shown in Figs. 2-26 in Volume 1 and 1-1 of the present volume.

In many oxidation-reduction reactions, especially ones involving organic compounds in aqueous media, negative charges combine with H^+ ions, so that the oxidation states A^- and AH are equivalent:

$$A^- + H^+ \longrightarrow AH \qquad (1\text{-}2)$$

One could say that an acid-base reaction is merged with an oxidation-reduction process, and electrons are transferred essentially as H atoms. The reactions

$$A^- \rightleftharpoons A + e^-, \ AH \rightleftharpoons A + e^- + H^+, \text{ and } AH \rightleftharpoons A + H \qquad (1\text{-}3)$$

are the same except for the expressed or implied involvement of H^+ ions. For many organic compounds the stable reduced and oxidized forms differ by two equivalents (electrons), and the reaction should be written

$$AH \rightleftharpoons A^+ + 2e^- + H^+ \tag{1-4}$$

in some cases and

$$AH_2 \rightleftharpoons A + 2e^- + 2H^+ \tag{1-5}$$

in others. The last of these corresponds to the formalism

$$H_2A \longrightarrow A + 2(H) \tag{1-6}$$

which will appear in the next section.

The involvement of H^+ ions in oxidoreduction is shown experimentally by the way in which pH affects the reaction. Reactions (1-4) and (1-5) are driven to the left at low pH (high H^+ concentration) in accordance with the mass-action principle.

1-2. Photosynthesis as a Photochemical Oxidation-Reduction Process

Green plants can use light energy to make carbohydrates from carbon dioxide and water, with oxygen evolved as a byproduct. The universal presence of Chl suggested decades ago that this pigment plays an essential photochemical role. The overall reaction is

$$6CO_2 + 6H_2O \xrightarrow[\text{(light, Chl)}]{} C_6H_{12}O_6 + 6O_2 \tag{1-7}$$

when a simple sugar such as glucose, $C_6H_{12}O_6$, is formed. More generally we can use (CH_2O) as the formula for one-sixth of a sugar molecule or an equivalent amount of any other carbohydrate and write

$$CO_2 + H_2O \xrightarrow[\text{(light, Chl)}]{} (CH_2O) + O_2 \tag{1-8}$$

Prior to the studies of C. B. van Niel, which began to have a profound effect on thinking about photosynthesis in the 1930s, little was known about the mechanism beyond what is expressed in Reaction (1-8). The absence of specific knowledge permitted a simple (but incorrect) picture, more or less as follows: Water plus carbon dioxide yields carbonic acid, H_2CO_3, which is bound in some way to a molecule of Chl. When the Chl molecule absorbs a quantum of light, the energy somehow causes a rearrangement of the carbonic acid to a "formaldehyde-peroxide," HO-O-CHO, which then dissociates into a formaldehyde-like substance and molecular oxygen. The oxygen is liberated, and the "photo-formaldehyde," (CH_2O), is built up to become carbohydrate. Unfortunately, no one could show the presence, in illuminated plants, of anything like formaldehyde. This appealingly simple picture of photosynthesis did not seem to lead in any fruitful direction, and was eventually abandoned by most investigators. Meanwhile, van Niel was developing an entirely different view of the process through his comparative studies of photosynthesis in green plants and in certain photosynthetic bacteria (Fig. 1-15).

Photosynthetic bacteria, like green plants, can grow at the expense of light energy. The essential pigment, bacteriochlorophyll (BChl), differs slightly from the green plant chlorophylls. Unlike the green plants, the bacteria do not evolve oxygen in their photosynthesis. Furthermore their ability to assimilate carbon dioxide during photosynthetic growth is dependent on the presence of a suitable oxidizable substance such as hydrogen, hydrogen sulfide, or any of a variety of simple organic compounds. The oxidation of this substance (called the *substrate*) balances the reduction of carbon dioxide; a typical overall reaction for bacterial photosynthesis can be written

$$CO_2 + 2H_2A \xrightarrow[\text{(light, BChl)}]{} (CH_2O) + H_2O + 2A \qquad (1\text{-}9)$$

where H_2A is a general representation of the oxidizable substrate and A is the product of its oxidation.

Some specific examples of H_2A in Reaction (1-9) are:

1. H_2A is hydrogen sulfide and A is sulfur.
2. H_2A is hydrogen and A is nothing.
3. H_2A is isopropanol, $CH_3CHOHCH_3$, and A is acetone, CH_3COCH_3.

Following the pattern suggested by these cases of bacterial photosynthesis, van Niel proposed that the photosynthesis of green plants is an analogous process in which H_2A is water and A is oxygen. To be consistent with Reaction (1-9), Reaction (1-8) should then be rewritten with one more H_2O on each side:

$$CO_2 + 2H_2O \xrightarrow[\text{(light, Chl)}]{} (CH_2O) + H_2O + O_2 \qquad (1\text{-}10)$$

This view immediately placed photosynthesis in a new light. The process consists of two parallel reactions, oxidation of the substrate H_2A and reduction of CO_2, which are initiated photochemically.

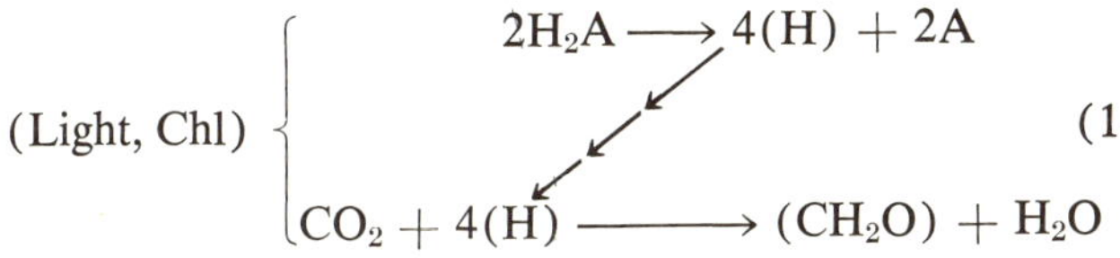
(1-11)

Pursuing his comparative observations further, van Niel noted that there are nonphotosynthetic bacteria that can assimilate CO_2 while oxidizing substrates in a reaction with O_2. The combustion of substrate with O_2 (respiration) provides energy for the reduction of CO_2. In fact some photosynthetic bacteria can live in this way, growing in the dark in the presence of O_2 by a respiratory metabolism as well as in the light by photosynthesis and consuming the same (organic) substrates in the dark as in the light. Taken together, these observations suggested to van Niel that neither the reduction of CO_2 nor the oxidation of substrate is exclusively the property of photosynthesis. The one essentially photosynthetic step is a photochemical reaction, sensitized by Chl, which forms separated reducing and oxidizing entities to drive the two metabolic processes. For convenience of visualization van Niel wrote a formalism in which water is split, in the primary photochemistry, into reducing and oxidizing fragments symbolized by (H) and (OH):

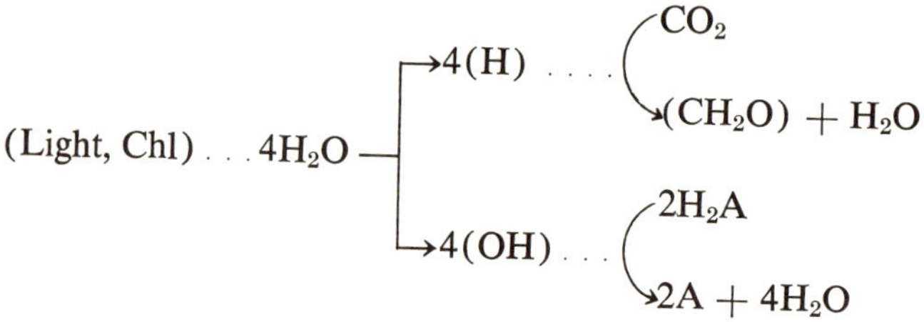

(1-12)

One can, however, avoid implying any involvement of water in the primary process and also omit the details of balanced chemical reactions, writing simply

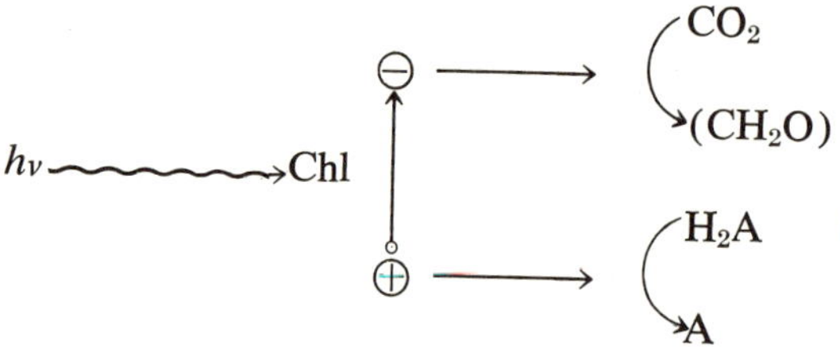

(1-13)

where $\ominus$ represents an electron (one reducing equivalent), and $\oplus$ is an electron vacancy, or oxidizing equivalent. Several points should be kept in mind in connection with this picture. First, the $\ominus$ and $\oplus$ actually reside on molecules that serve the functions of photochemical electron acceptor and donor, as indicated in the photochemical model shown in Fig. 1-1. Second, the involvement of H^+ and H, as needed to preserve a detailed chemical balance, has been left out of the picture merely for simplicity. Third, it actually takes four reducing equivalents to convert one molecule of CO_2 to carbohydrate, and correspondingly four oxidizing equivalents to liberate one molecule of O_2 from water (or, more generally, to oxidize $2H_2A$ to 2A). Fourth, the primary photochemical act provides no more than one ⊖ and one ⊕ for each light quantum.

Reaction (1-13) shows the essence of van Niel's revolutionary contribution to our thinking about photosynthesis. It indicates that the only reaction peculiar to photosynthesis is an oxidation-reduction driven by Chl in a photoexcited state. It should be added that green plant photosynthesis is the only known metabolic process in which H_2O serves as an oxidizable substrate, as a source of electrons for the reduction of CO_2 to carbohydrate and other cell materials.

In recent years the oxidoreductive nature of photosynthesis has been established beyond question through experiments (Secs. 1-5 and 1-9) showing the light-induced reduction and oxidation of substances including Chl or BChl. Earlier support for van Niel's picture came from two major quarters: On the one hand R. Hill showed that green leaf extracts could perform a light-dependent evolution of oxygen in which artificial electron acceptors, rather than CO_2, became reduced. Suitable electron acceptors, or *Hill reagents,* cover a wide range [1] including ferricyanide and benzoquinone. This showed that the assimilation of CO_2 could be separated from the photosynthetic process, leaving the evolution of O_2 intact. On the other hand H. Gaffron showed that some green algae could be adapted, by incubating them in the dark under hydrogen, to perform a bacterial type of photosynthesis. The algae would then assimilate CO_2 in the light while oxidizing H_2 as substrate, with little or no evolution of O_2. The algae reverted to "normal" green plant photosynthesis upon exposure to strong light. Apparently the adaptation involves activation of a hydrogenase enzyme that allows H_2 to compete with H_2O as oxidizable substrate for photosynthesis. This demonstration shows that O_2 evolution can also be deleted while leaving the rest of the photosynthetic process apparently intact.

[1] The dye 2,6-dichlorophenol indophenol is blue in its oxidized form and colorless when reduced. The rapid decolorization of the blue form by illuminated leaf extracts (buffered to a pH near 7) provides a simple and striking demonstration of the reducing power generated in photosynthesis. Oxygen evolution is best observed in a submerged aquatic plant exposed to bright light.

Despite the results with hydrogen-adapted algae, Gaffron criticized van Niel's scheme strongly on other grounds. In many cases of bacterial photosynthesis an organic substrate is consumed, and a product is stored within the bacteria, with little difference in oxidation-reduction level between the substrate and the storage product. A particularly striking case is the assimilation of β-hydroxybutyric acid, $CH_3CHOHCH_2COOH$, which becomes stored as a polymer of the same substance in a photosynthetic (light-dependent) process. All that is required of photosynthesis in this case is the small amount of energy needed to join the molecules of the acid together as a polymer. It would seem stilted and unnatural to insist that the acid, serving as substrate H_2A, is degraded to the level of CO_2 and the CO_2 is then transformed into a polymer of the same acid. A simpler picture, suggested by Gaffron, is that the photosynthetic process feeds a reservoir of chemical energy which drives the more or less direct assimilation of organic substrates:

$$h\nu \rightsquigarrow \text{BChl} \longrightarrow \text{energy} \left(\begin{array}{l} \nearrow \text{storage products} \\ \searrow \text{substrates} \end{array} \right. \qquad (1\text{-}14)$$

This picture could in principle be extended to describe all cases of bacterial photosynthesis.

While these alternatives were being considered, O. Meyerof, F. Lipmann, and others were clarifying the biochemical patterns of energy storage and discovering the central role of ATP. We shall return to the problem of reconciling van Niel's and Gaffron's viewpoints after a digression into the biochemistry of ATP.

1-3. Adenosine Triphosphate

ATP is composed of one molecule of the base adenine, one of the five-carbon sugar ribose, and three of ortho-phosphoric acid, as

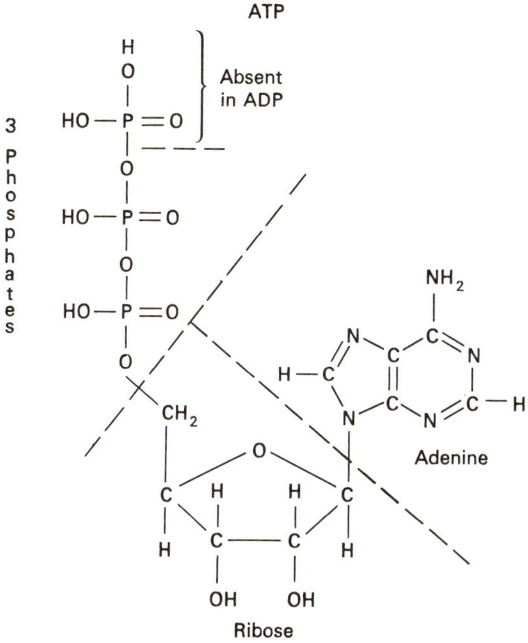

Fig. 1-2. The chemical structure of adenosine triphosphate (ATP).

shown in Fig. 1-2.[1] The same thing with only two phosphoric acid residues is adenosine diphosphate (ADP). Abbreviating orthophosphoric acid as P_i (for inorganic phosphate) we can write the reaction

$$ADP + P_i + 8\ \text{kcal} \rightleftharpoons ATP + H_2O \qquad (1\text{-}15)$$

The energy difference is about 8 kcal per mole, or about 0.3 electron volt per molecule under physiological conditions; it varies with pH, temperature, and other factors. To show this reaction in more

[1] A related molecule of biological importance, NADP, is shown in Fig. 1-3, and the structures of some chlorophylls are shown in Fig. 1-4. These last two figures will not concern us in this section.

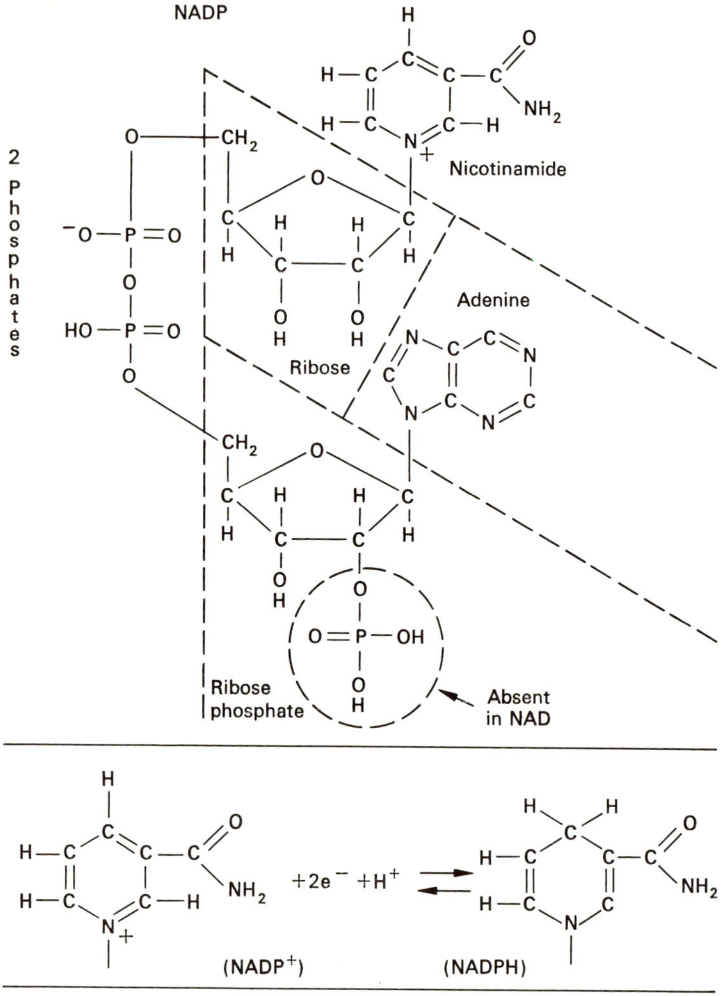

Fig. 1-3. The chemical structure of nicotinamide–adenine dinucleotide phosphate, NADP. Without the "odd" phosphate we have NAD. Note the similarity to ATP. Interconversion between the oxidized form NADP^{+} *and the reduced form NADPH, which takes place on the nicotinamide part, is shown below.*

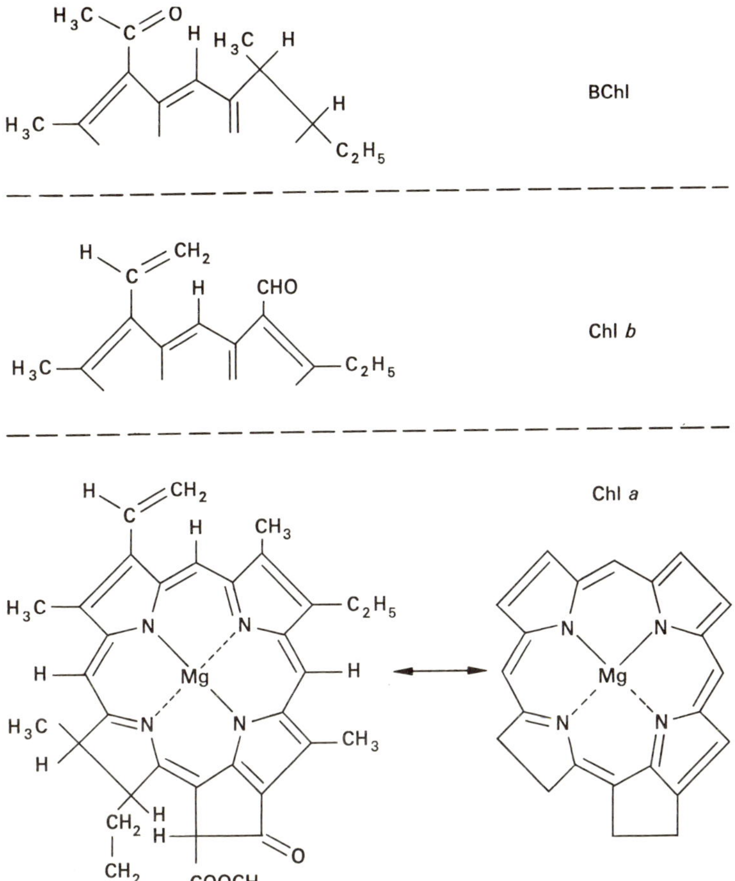

Fig. 1-4. The chemical structures of chlorophylls a *and* b *and bacteriochlorophyll (BChl). The presence of a carbon atom is implied at each unlabeled junction of bonds. In Chl* a *and* b, *but not in BChl, the pattern of alternating single and double bonds is in resonance with the one sketched at the right. The residue R is a long-chain hydrocarbon,* $C_{20}H_{39}$, *or phytyl, in Chl* a *and* b *and something similar in BChl.*

detail we can represent all but the terminal phosphate of ATP by the letter R and write [1]

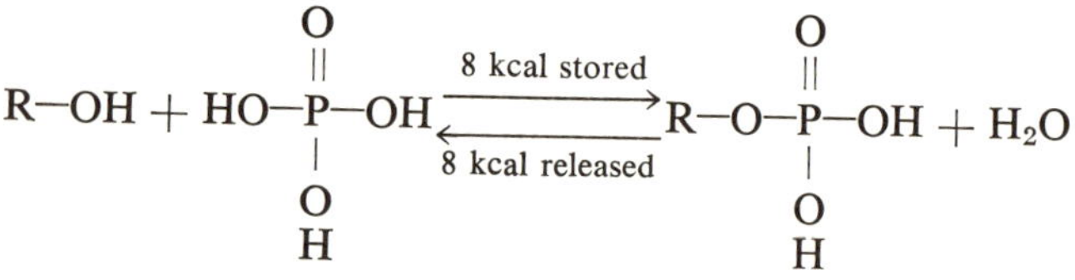

(1-16)

ADP ATP

The anhydride bond that connects ADP to phosphate is often written as ~, so that ATP can be written ADP ~ P. More cryptically, and more generally, the addition of phosphate to an organic molecule through the formation of an anhydride bond can be written

$$X + P_i \longrightarrow X \sim P$$

or simply

$$P_i \longrightarrow \sim P$$

The storage and release of energy shown in Reaction (1-16) is useful, because it can be coupled to other metabolic processes that yield or require energy. The pool of ATP therefore acts as a clearinghouse and switching station for energy in life processes.

An energy-yielding process that is important in animals and also occurs in plants is respiration, the combustion of substrates with oxygen. This process involves a chain of linked oxidation-reduction reactions through an *electron transport system* as shown schematically in Fig. 1-5. Each step of this sequence is a reaction between a reducing agent and an oxidizing agent (for example, between YH_2 and Z in Fig. 1-5), proceeding in the direction favored by the downhill flow of energy. Thus the energy available from the overall combustion is given up in small increments in the succes-

[1] At a pH near 7, ADP, designated R-OH in Reaction (1-16), is mainly dissociated into $R\text{-}O^-$ and H^+. Phosphoric acid and ATP are similarly ionized.

sive steps. Some of these steps are coupled to reactions of the type $P_i \longrightarrow \sim P$, so that most of the energy can be conserved in compounds of the form $X \sim P$, including ATP. This process, by which the energy of respiration is stored in "high-energy" phosphate compounds, is called *oxidative phosphorylation.* The mechanism of coupling is poorly understood; it will be discussed in Sec. 1-14.

A basic pair of events in oxidative phosphorylation is outlined in Fig. 1-6*a*. The reaction $YH_2 + Z \longrightarrow Y + ZH_2$ liberates

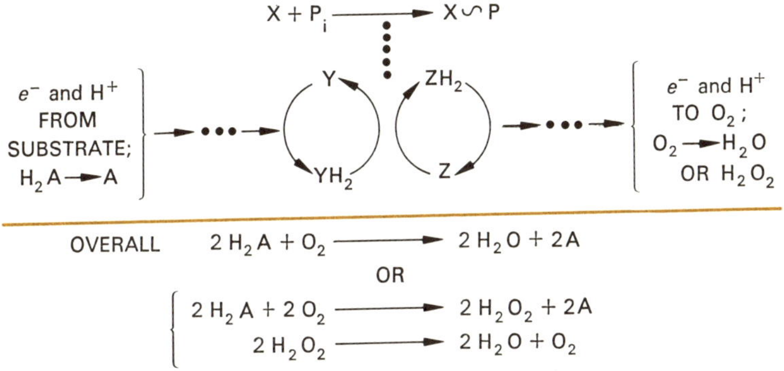

Fig. 1-5. Showing the flow of reducing equivalents from substrate to oxygen during respiration, through a sequence of electron carriers, and the coupled incorporation of inorganic phosphate (P_i) *into "high-energy" compounds* ($X \sim P$).

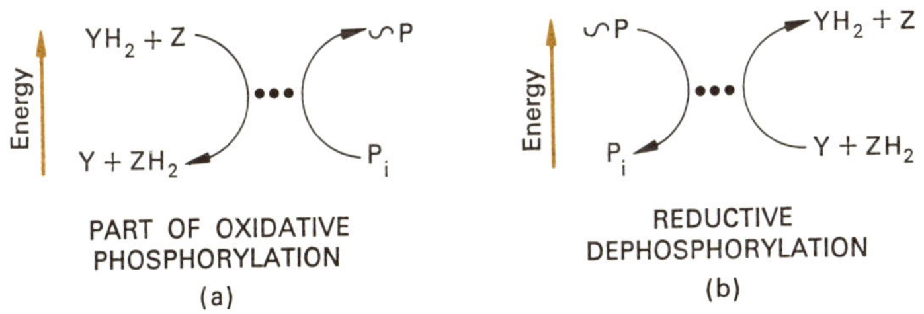

Fig. 1-6. Compare Fig. 1-5. (*a*) *Energy made available from an oxidation-reduction reaction can be stored as a high-energy phosphate compound* ($\sim P$). (*b*) *In a reverse process, phosphate bond energy can drive an oxidoreduction uphill, against the energy gradient. In this way a stronger reductant,* YH_2, *can be formed at the expense of a weaker one,* ZH_2.

energy, some of which is captured in the reaction $P_i \longrightarrow \sim P$. Coupled reactions of this kind have been shown to be reversible, as indicated in Fig. 1-6*b*. The phosphate bond energy of $\sim P$ can be released in the reaction $\sim P \longrightarrow P_i$, and this energy can drive the oxidation-reduction process uphill, against the energy gradient: $Y + ZH_2 \longrightarrow YH_2 + Z$. In this picture, YH_2 is a stronger reducing agent than ZH_2. The process of Fig. 1-6*b* can therefore be called a reductive dephosphorylation.

The possibility of reductive dephosphorylation increases the flexibility of metabolic patterns in both photosynthetic and respiring organisms, as we shall see.

In plants and photosynthetic bacteria there are electron transfer sequences that begin with the primary photochemical act, as implied in Reaction (1-13). One can expect, in analogy with oxidative phosphorylation, that some steps in these sequences can be coupled to the formation of ATP and other high-energy phosphate compounds. Such "photosynthetic phosphorylation" was thus predicted and demonstrated by D. I. Arnon, A. W. Frenkel, and others in the 1950s.

1-4. Patterns of Electron Transport in Photosynthetic Bacteria and Green Plants

When van Niel developed his formulation it seemed that any recombination of the primary (H) and (OH), or $\ominus$ and $\oplus$, would be wasteful. The function of H_2A appeared to be the removal of the $\oplus$ so as to avoid a reaction between the $\ominus$ and the $\oplus$, and allow the $\ominus$ to bring about the reduction of CO_2. But with the discovery of phosphorylation coupled to electron transport, recombination of the primary entities could be visualized as a useful process:

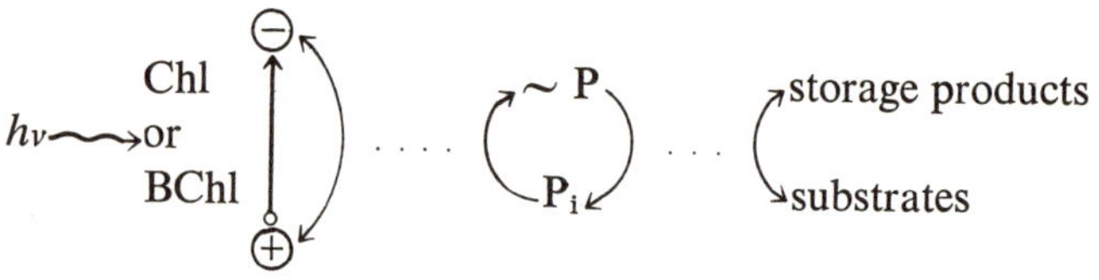

(1-17)

In fact, this version of Gaffron's viewpoint [Reaction (1-14)] could in principle be superimposed on van Niel's scheme [Reaction (1-13)], with the primary ⊖ and ⊕ capable of following either the "cyclic" (1-17) or the "open" (1-13) pathways. The final result of either pathway is the oxidation of substrate and the synthesis of cell materials. The cyclic and open mechanisms cannot be distinguished on the basis of the overall chemistry; any distinction requires the study of phosphorylated intermediates including ATP.

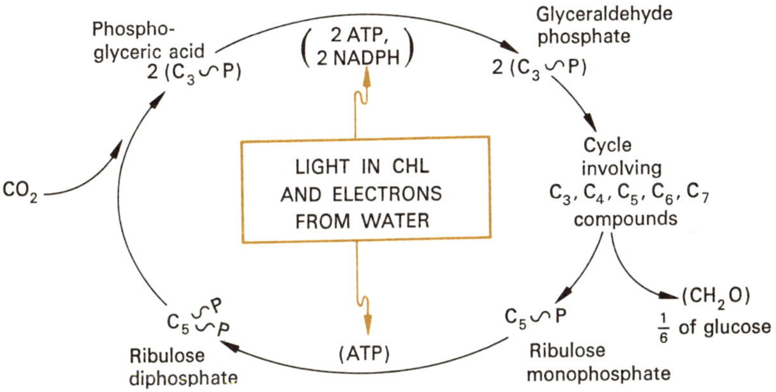

Fig. 1-7. The Calvin-Benson cycle for photosynthetic carbon dioxide assimilation. The light reactions furnish the necessary energy and reducing power in the form of ATP and NADPH.

The actual requirements for the conversion of CO_2 to sugar have now been shown to include both stable reductant *and* ATP. A major pathway elucidated by M. Calvin and his associates, outlined in Fig. 1-7, utilizes two molecules of NADPH (see the lower part of Fig. 1-3) and three of ATP for each molecule of CO_2 assimilated. The incorporation of CO_2 occurs in a reaction with the doubly phosphorylated five-carbon sugar ribulose diphosphate, forming two molecules of the three-carbon compound phosphoglyceric acid. This is reduced to glyceraldehyde phosphate in a reaction that draws upon the energy of ATP and the reducing power of NADPH as indicated in Fig. 1-7 and in the following reaction outline:

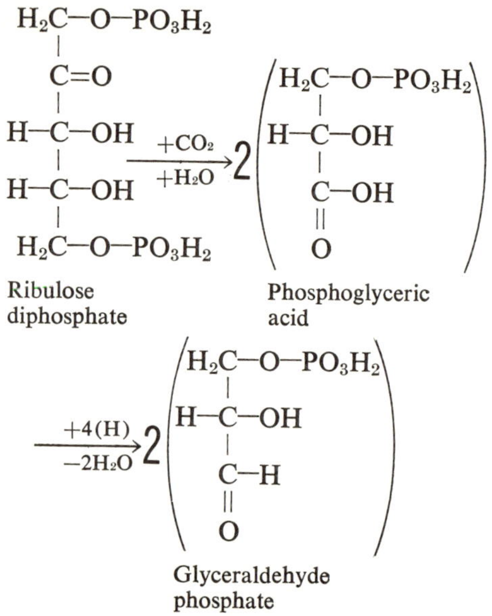

(1-18)

The two molecules of glyceraldehyde phosphate (six carbon atoms in all) then enter into a complex cycle involving three-, four-, five-, six-, and seven-carbon compounds, and eventually one carbon atom enters the structure of glucose, and the other five appear as ribulose monophosphate. The cycle is completed when ribulose monophosphate is converted to the diphosphate at the expense of ATP. The function of light in this process is to provide the NADPH and ATP needed for the conversion of phosphoglyceric acid to glyceraldehyde phosphate and of ribulose monophosphate to the diphosphate. The actual incorporation of CO_2 does not require light; it requires only the presence of ribulose diphosphate.

In view of these findings the combination of Reactions (1-17) and (1-14) should be written as indicated in Fig. 1-8. A scheme of this sort has enough flexibility to allow the plant to make reducing power and ATP in whatever proportions are needed.

We must now anticipate a fact that will be documented in Sec. 1-9: that photosynthesis in green plants involves the operation

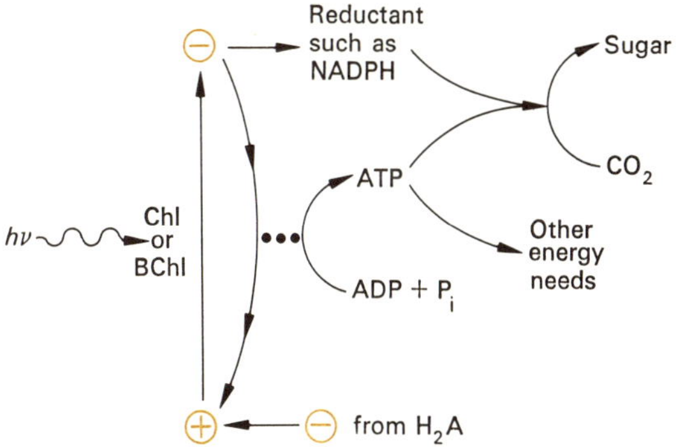

Fig. 1-8. Some possible patterns of electron flow and phosphorylation in photosynthesis. This is by no means a complete description of the process; see Figs. 1-9 and 1-17.

of two distinct photochemical systems. One of these (system II) makes a weak reductant and an oxidant strong enough to oxidize water and thereby liberate oxygen. The other (system I) makes a weak oxidant and a reductant capable of reducing NADP (Fig. 1-3) and thereby furnishing the reducing power to convert CO_2 to sugar. The two systems interact through a chain of electron carriers as sketched in Fig. 1-9. In this figure the vertical direction is a scale of energy, indicated as oxidation-reduction potential. A large negative value corresponds to an electron of high energy, that is, a strong reductant.

The energy scale in Fig. 1-9 shows that the total span between the strong reductant made by system I and the strong oxidant made by system II is about 1.4 electron volts per molecule. This could be provided by a single quantum absorbed by Chl, since one quantum at about 670 nm has an energy of about 1.8 electron volts, but in fact the plant uses two photochemical steps, as indicated.

The "downhill" flow of an electron from the top of system II to the bottom of system I releases about 0.5 electron volt, enough energy to make one molecule of ATP from ADP and P_i. Then two quanta, one in each photochemical system, could promote an elec-

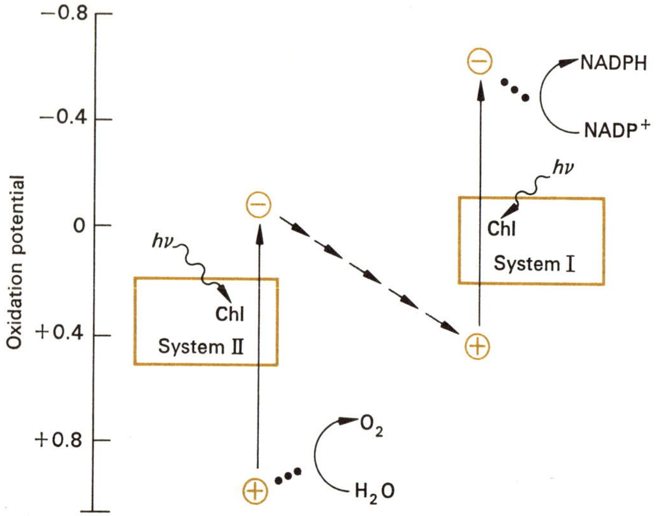

Fig. 1-9. Outline of a representation of green plant photosynthesis, showing the interaction of two photochemical systems through a chain of electron carriers. The vertical scale indicates electron energy (oxidation-reduction potential). A large negative value signifies high electron energy or strong reducing power.

tron through the entire chain and at the same time make a molecule of ATP. Eight quanta could then transfer four electrons through the entire sequence, forming two molecules of NADPH (see Fig. 1-3, bottom) and four of ATP. This would be enough for the assimilation of one CO_2, with one ATP left over. In actuality, however, this process (called noncyclic phosphorylation) yields only about one ATP for every two electrons passing through the sequence. Moreover, there are cyclic pathways by which ATP can be made without the storage of strong reductant. For example (and this is the only case that has been verified), the flow of electrons from the "top" of system I to a point between systems II and I can complete a cycle which is driven by system I and which is coupled to phosphorylation as in Reaction (1-17). This "cyclic phosphorylation" uses in part the same electron carriers, between systems I and II, that function in the noncyclic mode. We do not know if the cyclic system uses one or more sites for the coupling of electron flow to phosphorylation that are distinct from the site(s) employed in the noncyclic process.

A cyclic path is included in the more detailed version of Fig. 1-9 that is shown in Fig. 1-17. There is no experimental justification for further elaboration of these electron flow patterns, but more complicated networks are certainly possible. For example, an electron at any point in the system might be promoted to a higher level by reductive dephosphorylation, and then reenter the pathways. The reader can easily imagine a variety of possible consequences. In any event, what we know about green plant photosynthesis suggests that about nine quanta should suffice for the assimilation of one CO_2, and this is consistent with measurements of the quantum efficiency.

The most conspicuous biochemical difference between green plants and photosynthetic bacteria is the inability of the latter to evolve oxygen, coupled with their dependence on oxidizable substrates other than water. It has therefore seemed natural to regard the photochemical system of the bacteria as analogous to green plant system I, with electrons from substrate taking the place of electrons from system II. This flow of electrons from substrate would play a part only when the system operates in a noncyclic, or open-ended, manner. The bacteria can also perform cyclic phosphorylation, in which case the metabolism of substrate is not tied directly to the light-driven electron flow.

Actually the noncyclic mode of bacterial photosynthesis has not been easy to demonstrate and remains to be proved. The bacteria appear to use NADH as a strong reductant, in place of the NADPH used by green plants. However, all efforts to observe light-dependent formation of NADH (or any other "strong" reductant) in the bacteria have been disappointing in their quantitative aspects. Furthermore, the formation of NADH is dependent on the capacity for phosphorylation, suggesting reductive dephosphorylation as the mechanism. Finally, there are ways (see Sec. 1-11) to estimate the oxidation-reduction potentials associated with the photochemical acts; a potential more reducing than about

—0.35 volt is needed for the direct formation of NADH or NADPH. The potential at the reducing side (the top) of green plant system I is about —0.6 volt, but that of the bacterial system appears to be limited to about —0.1 volt. It is therefore probable that the photochemistry of the bacteria is used mainly for cyclic phosphorylation. This view, an outgrowth of Gaffron's position relative to van Niel's, has been advanced especially by H. Gest and R. Y. Stanier.

We shall attend later to some very recent evidence that the bacteria may contain two kinds of photosystems, one for cyclic and the other for noncyclic metabolism, perhaps the evolutionary precursors of green plant systems I and II.

PHOTOSYNTHETIC UNITS: LIGHT-HARVESTING SYSTEMS AND REACTION CENTERS

1-5. Photochemical Reaction Centers

The foregoing biochemical events are set in motion at photochemical reaction centers. In all known cases these are of the sort shown in Fig. 1-10 (see also Fig. 2-27*a* and Sec. 2-13 in Volume 1), where an electron is transferred from the sensitizing pigment to an acceptor, producing oxidized sensitizer and reduced acceptor. The identification of reaction centers began with L. N. M. Duysens' observation in 1953 that when photosynthetic bacteria are illuminated, the intensity of the long-wave absorption band due to BChl is diminished slightly. The original absorption intensity returns rapidly (usually in less than a second) in the dark. This reversible bleaching is due to the oxidation of a small fraction of the total BChl. The major part of the BChl acts only as a light-harvesting antenna, whereas a small fraction is specialized to act as sensitizer for the kind of photochemistry shown in Fig. 1-10.

Using suspensions of broken cells of photosynthetic bacteria, called *chromatophore suspensions,*[1] we have found ways to remove

[1] Chromatophores are fragments of the cell membrane of photosynthetic bacteria. This membrane carries the photochemical apparatus for photosynthesis, as well as the light-harvesting apparatus. Disruption of the cells yields functional fragments in which the membrane forms a closed surface,

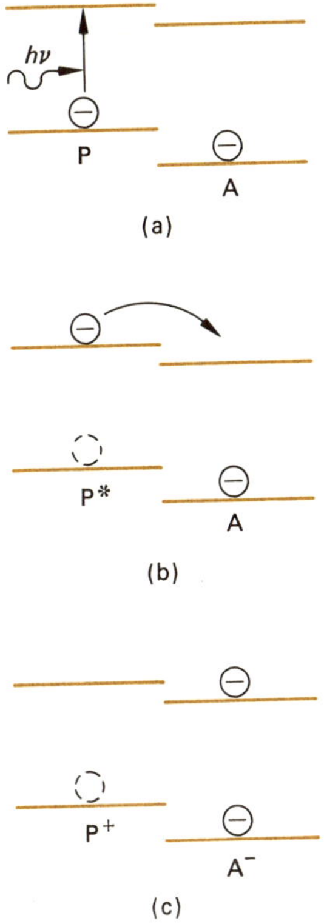

Fig. 1-10. In all known photosynthetic reaction centers the sensitizing pigment is a specialized Chl or BChl, called P. In the light reaction P becomes oxidized and thereby appears as the primary oxidizing entity. At the same time an acceptor A becomes reduced.

selectively nearly all the light-harvesting BChl and thus to make preparations that are greatly enriched in components of the reaction centers. In these reaction center preparations the light-induced

separating an inner phase from the external medium. The fragments are usually about 0.1 micron (1000 Å) in diameter.

bleaching (due to oxidation) of the specialized BChl is conspicuous and easy to study.

In the photosynthetic bacterium *Rhodopseudomonas spheroides,* the photochemically active component of BChl has been named P870 (P for pigment and 870 for the wavelength of maximum absorption). The reversible light-induced bleaching of this pigment is shown in Fig. 1-11. For comparison the phenomenon is shown both for a suspension of chromatophores, in which most of the absorption at 870 nm is due to the light-harvesting BChl, and for a reaction center preparation, in which all the 870-nm absorption is due to P870. The dashed curve represents the absorption spectrum measured under strong illumination; the same spectrum can be produced by adding an oxidizing agent (such as potassium ferricyanide) to the material rather than illuminating it.

It can be seen in Fig. 1-11 that a component absorbing at 800 nm (called P800) exhibits a shift of the absorption band to shorter wavelengths when P870 is bleached. We do not know if P800 has any functional significance. Chemically, P800 and P870 are molecules of BChl in an environment specialized for photochemistry. The environment determines the positions of the absorption maxima (800 and 870 nm respectively) and endows P870 with the property of photochemical electron donor. Both P800 and P870 are extracted from reaction center preparations when methanol is added; in the methanolic solution these pigments are indistinguishable from light-harvesting BChl that has been extracted in the same way from cells or chromatophores. In methanol the long-wave absorption maximum is at 770 nm.

The simplest interpretation of this situation is that P870 is specialized by virtue of its proximity to a suitable electron acceptor and P800 is merely the nearest neighbor of P870. The band shift of P800 may then be an inconsequential response to the oxidation of P870.

The quantum efficiency for photochemical oxidation of P870 is greater than 70 percent, both in cells, where most of the light is absorbed by the light-harvesting BChl, and in reaction center preparations, where the light is absorbed by P800 or by P870 itself.

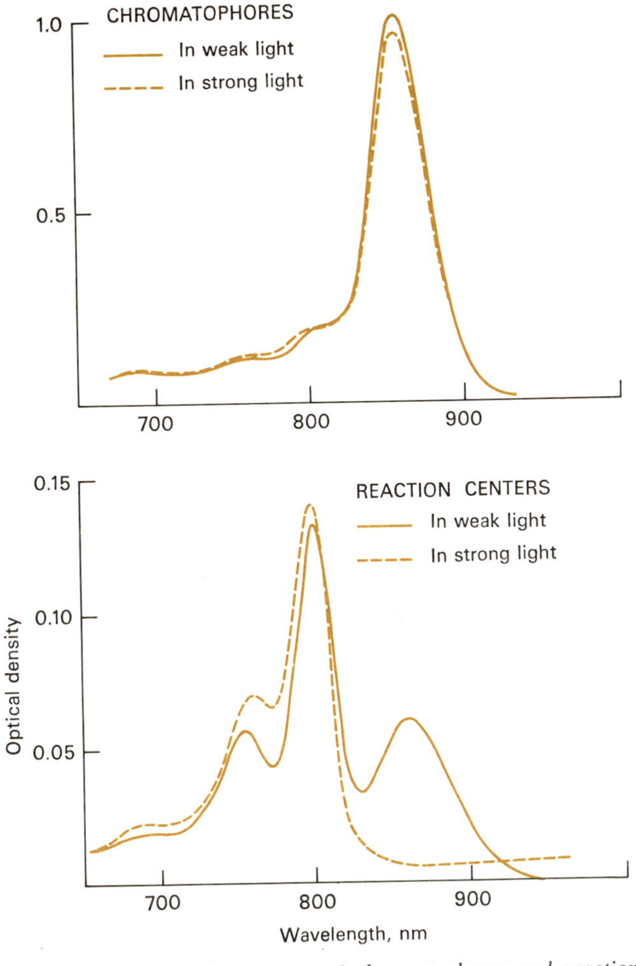

Fig. 1-11. Absorption spectra of chromatophores and reaction centers prepared from the photosynthetic bacterium Rhodopseudomonas spheroides. *Solid curves, spectra measured in weak light; dashed curves, strong light superimposed in order to bleach the reaction center pigment P870. Chromatophores are fragments of the cell membrane. They contain the entire photochemical apparatus including the light-harvesting BChl, which accounts for most of the absorption at 870 nm. Reaction centers are detached from these chromatophores by adding a detergent and then isolated by centrifuging the mixture in a density gradient. Preparations made by the author.*

W. R. Sistrom has developed mutant strains of *R. spheroides* which have a normal complement of light-harvesting pigments but which show no evidence for the presence of P870 or P800. These strains seem incapable of performing any useful photochemistry. They cannot grow photosynthetically; they must be maintained as typical aerobic bacteria, growing in the dark at the expense of energy from respiration.

Both the high quantum efficiency of P870 oxidation and the incapacity of cells lacking P870 testify to the importance of this pigment for photosynthesis.

Pigments analogous to P870 have been found in all kinds of photosynthetic organisms and named according to the long-wave absorption maximum. Algae and green plants contain P700, which is probably Chl *a* (see Fig. 1-4) in the specialized context of a reaction center. P700, first noted and studied by B. Kok and by H. T. Witt, appears to be the primary photochemical electron donor for photochemical system I in green plant photosynthesis (see Figs. 1-9 and 1-17).

The photochemical electron acceptor has not yet been identified for any photosynthetic system and has been called variously X, A, Q, E, etc., by investigators dealing with different systems.

Excitation and the consequent electron transfer in a reaction center can be written

$$P \cdot A \xrightarrow[\text{(light)}]{} P^* \cdot A \longrightarrow P^+ \cdot A^- \qquad (1\text{-}19)$$

where P represents P700, P870, etc., and A is the electron acceptor. P^+ and A^- are the primary photochemical products; they must be restored to the forms P and A before the reaction can happen again. Normally this restoration is not by a direct back-reaction ($P^+ \cdot A^- \longrightarrow P \cdot A$), which would be wasteful.[1] Instead, it is by either a cyclic or a noncyclic process as indicated in Reactions

[1] The wasteful back-reaction is avoided because the processes that initiate useful electron flow are much faster or more probable.

(1-17) and (1-13) and Fig. 1-8, with the ⊕ and the ⊖ identified as P^+ and A^-. Depending on the relative rates of the two restoring events ($P^+ + e^- \longrightarrow P$ and $A^- \longrightarrow A + e^-$) the species $P^+ \cdot A$ or $P \cdot A^-$ may appear as transitory intermediates in the operating cycle of photochemistry and restoration. In any case the reaction center must be restored fully, to the form $P \cdot A$, before it can work again.

We can expect, of course, that chemical systems beyond the reaction centers also go through consecutive cycles of activity and recovery. The time needed for recovery of all components of the photosynthetic system can be measured by assaying the fruits of consecutive flashes of light. Suppose that a leaf is exposed to a flash so brief and so strong that each reaction center is made to react just once. When the leaf has dealt completely with the products of this flash, it can make full use of another flash.[1] Detailed experiments of this kind were first reported in 1932 by R. Emerson and W. Arnold, who studied oxygen evolution in suspensions of the green alga *Chlorella* exposed to a series of light flashes. Each flash lasted about 10^{-5} sec. They found that the yield of oxygen per flash was maximal if the interval between flashes was greater than about 0.04 sec; this was therefore the time needed by the algae to dispose of the products of one flash and to be ready to utilize the next flash with full efficiency. Thus the "dark" chemical machinery could keep pace with about 25 cycles of operation per second in the reaction centers. Similar results have been obtained with photosynthetic bacteria, using a variety of criteria. Measurements of the bleaching and recovery of P870, or of the oxidation and reduction of a cytochrome (see Fig. 3-11, Volume 1) that donates electrons to oxidized P870, or of the capacity for ATP formation in intermittent light all show that reaction center turnover rates of about 25 to 100/sec can be maintained.

[1] This simple reasoning can be confounded in a variety of ways. For example, the evolution of oxygen and the assimilation of carbon dioxide both involve pools of chemical reserves, the concentrations of which should depend on the prior history of the system. The ability to use the products of one flash can thus depend in surprising ways on the input of earlier flashes.

1-6. The Light-harvesting Antenna

Imagine a primordial photosynthetic cell in which the only Chl (or BChl) is the component associated with electron acceptors and donors:

A
·
Chl
·
D

There are no accessory light-harvesting pigments. Suppose that this cell is growing in moderate or shady daylight. Then by the methods of Probs. 9 and 10 in Volume 1 it can be estimated [1] that, on the average, each Chl molecule absorbs about 1 quantum/sec. In view of a turnover capability of 25 to 100/sec for the "dark" machinery, the system is being grossly underdriven. Now imagine a mutation in which the photoactive Chl at each reaction center is surrounded by some additional Chl molecules:

A
Chl· . · . ·Chl
Chl
Chl· · · ·Chl
D

The extra molecules of Chl are close enough to the photoactive one that they form a common light-harvesting system. Quanta absorbed in any part of the system have a high probability of being used for photochemistry at the original photoactive one; see Sec.

[1] The estimate involves an integrated product of the incident light intensity and the extinction coefficient ε of the Chl over its entire absorption spectrum. Absorption occurs chiefly in the two main bands centered at about 430 and 670 nm.

2-17 and especially Fig. 2-33*c* in Volume 1. The mutated cell, being able to drive its dark machinery more nearly to capacity, should have a severalfold advantage (depending on the size of the Chl antenna) over its nonmutated neighbors in terms of growth rate. Its progeny should therefore come to predominate in the population.

One should expect the light-harvesting antenna to grow, by successive mutations and natural selection, until the dark systems are being driven close to their capacity under the prevailing light intensity. Too large an antenna would be a useless burden on the economy of the cell, of course. A variety of tests (see later) show that in most plants, algae, and photosynthetic bacteria there are about 50 to 500 molecules of light-harvesting Chl for each reaction center.

A photochemical reaction center and its share of light-harvesting pigment is called a *photosynthetic unit.* This may be a distinct structural entity in the cell, or it may be simply one fraction of an extended "ocean" of light-harvesting pigments dotted here and there with reaction centers. These alternatives are illustrated in Fig. 1-12, with the photoactive Chl designated P. The structure of the photosynthetic apparatus will be discussed briefly in Sec. 1-13.

The economics of the size of the light-harvesting antenna can

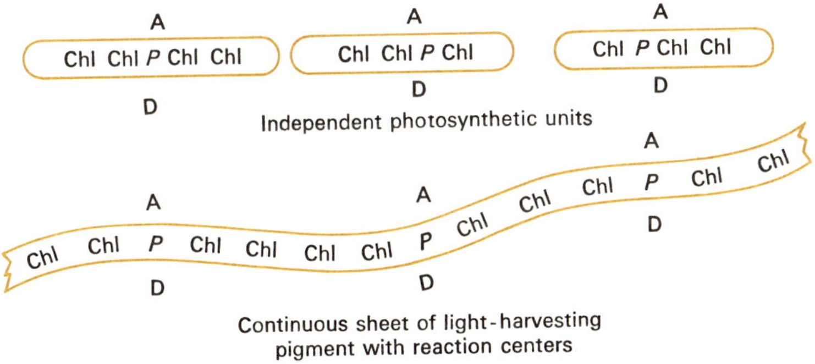

Fig. 1-12. Two models for the structure of the photosynthetic apparatus. The main functional difference is that in the case of independent units a quantum of energy cannot visit more than one reaction center.

be illustrated by a contemporary example. There is a Chl-deficient mutant of tobacco, studied by G. Schmid and H. Gaffron, in which the Chl content per unit of leaf area is about twenty times less than in the parent strain of the plant. In shaded daylight the growth of the mutant is severely curtailed, but in full sunlight this parchment-colored plant grows nicely. In fact, in sufficiently strong light the mutant can assimilate carbon dioxide *faster* than the parent strain, per unit of leaf area. Apparently the Chl-deficient mutant has an abnormally small light-harvesting antenna, but this deficiency can be overcome by using very strong light. The mutant, lacking a large antenna, actually has room for more reaction centers and dark chemical systems than the parent strain. Such antennaless mutants might be useful in desert or tropical agriculture, where full sunlight is common.

The existence of photosynthetic units, or of reaction centers served by light-gathering antennas, was first implied by the results of some of Emerson and Arnold's experiments on photosynthesis in flashing light. One of the experiments, dealing with photosynthetic oxygen evolution by *Chlorella* algae, was to give a succession of flashes spaced more than 0.04 sec apart (to allow full recovery after each flash) and intense enough to give the maximum possible yield of oxygen. Each flash was so brief that it allowed just one "turnover" of the photochemical system. The product of one flash was found to be never more than about 1 molecule of O_2 for every 3,000 molecules of Chl in the algae. Nevertheless the efficiency, using light absorbed by any of these Chl molecules, was about one O_2 evolved per 8 quanta absorbed. Eight quanta, absorbed anywhere in a set of 3,000 Chl molecules, could generate the chemistry needed for evolution of one molecule of O_2. If we imagine that 1 quantum drives one photochemical act, we may say that a single quantum absorbed in a set of about 3,000/8 (or about 400) Chl molecules causes just one oxidation-reduction event at a reaction center. This produces one-eighth of the wherewithal for the evolution of an O_2 molecule. This experiment therefore suggested that *Chlorella* cells contain photosynthetic units in which about 400 molecules of Chl serve one reaction center.

This implication of Emerson and Arnold's experiment has

been confirmed in several ways. Chemical phenomena associated with photosynthesis, including such "primary" effects as oxidation of P700 or P870 at the reaction centers, can be analyzed in terms of their dependence on light intensity and on time, so as to yield the effective size of the light-gathering unit that serves one reaction center. More directly, photosynthetic tissues can be analyzed for their content of P700, or P870, or other minor constituents such as cytochromes that participate in photosynthetic electron transport. The ratios of total Chl or BChl to these "special" molecules can be taken as indicative of the size of the photosynthetic unit. In green plants these approaches can give the ratios of light-harvesting pigments to reaction centers in systems I and II separately, through study of constituents and reactions peculiar to each photosystem (experimental separation of the reactions of these two photosystems will be described in Sec. 1-9).

Further details about the numbers and types of light-harvesting pigments that form photosynthetic units in various organisms will emerge in the next section.

1-7. The Diversification of Light-harvesting Systems

Having speculated about the evolution of the light-harvesting antenna, let us imagine how the diverse pigments of contemporary photosynthetic organisms might have evolved. As a point of departure we shall consider the photosynthetic bacteria that contain BChl as the main light-harvesting pigment. Purified BChl dissolved in an organic solvent shows principal absorption maxima near 375 and 770 nm and a smaller peak near 590 nm. The environment of BChl in the living cell shifts the long-wave absorption maximum to positions ranging from 800 to 890 nm in various species of bacteria.[1] Often there are three different environments for the BChl in a single cell, producing peaks at about 800, 850, and 870 to 890 nm. The photochemically active pigment, such as P870, has its

[1] These shifts result from changes in the wave functions of those electrons that are involved in the optical transition. Two possible causes are the interactions of these electrons with neighboring BChl molecules and distortions due to binding of the BChl to proteins and lipids in the cell membrane.

absorption maximum near that of the longest wavelength component of BChl in these bacteria.

For the purpose of discussion we can start arbitrarily with an ancestral photosynthetic bacterium whose light-harvesting BChl had a peak of absorption at 800 nm as well as the usual peaks at 375 and 590 nm. Consider the plight of a cell near the bottom of a dense population of such bacteria, illuminated from above. The wavelengths of light which would be especially useful for its growth are those at the peaks of absorption, but these are the very ones that have been absorbed most strongly by the overlying bacteria. The only wavelengths that filter through the population are those at which the cell absorbs poorly (minima in the absorption spectrum). Now suppose that through a random mutation the deprived cell brings about a change in the environment of some of the BChl in its progeny, causing the long-wave absorption peak to shift from 800 to 850 or 870 nm. These new wavelengths pass readily through the layer of 800 nm–absorbing cells and afford luxuriant growth of the new phenotype in competition with the old.

When the possibilities of this maneuver have been exhausted, another trick can be introduced: The absorption peak can be shifted by changing the BChl chemically rather than by altering its environment. This amounts to inventing a new light-harvesting pigment. An apparent example is found in the green photosynthetic bacteria. These creatures contain a small proportion of BChl absorbing at 810 nm, and a photochemically active P840 that is probably BChl also. But the preponderant light-harvesting pigment is chemically distinct from BChl. It is called *chlorobium chlorophyll* (CChl) and occurs, in different green bacteria, in two forms with absorption peaks at 725 and 750 nm respectively. Finally there are photosynthetic bacteria with a BChl *b,* chemically distinct from the usual BChl. The long-wave absorption maximum of BChl *b* is at 800 nm in organic solvents but is shifted to about 1,025 nm in cells of *Rhodopseudomonas viridis.* In these bacteria the reaction centers contain a P985 which is probably BChl *b* as well.

Figure 1-13 shows absorption spectra of chromatophores from various photosynthetic bacteria (if measurements are made with chromatophores rather than whole cells, the problem of scattering

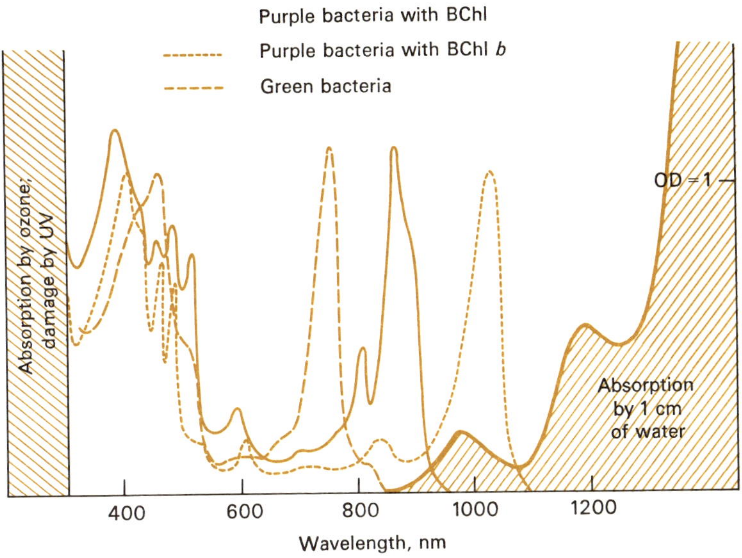

Fig. 1-13. Absorption spectra of chromatophore suspensions from three general types of photosynthetic bacteria (some variation occurs within each type). Absorption by water and ozone in the atmosphere and damaging effects of ultraviolet limit the usable parts of the spectrum. These limitations are indicated by the shaded areas.

is avoided). The figure also shows what parts of the spectrum (unshaded regions) are usable for photosynthesis. The infrared beyond about 1,100 nm is absorbed strongly by water in the atmosphere, and the ultraviolet below about 300 nm is attenuated by ozone in the upper atmosphere. If it were not for this absorption by ozone, all living things exposed to sunlight would have difficulty in surviving the damaging effects of the ultraviolet below 300 nm (see Chap. 5). It can be seen from Fig. 1-13 that the absorption bands of pigments in the photosynthetic bacteria cover most of the usable parts of the spectrum. Some of the peaks between 400 and 550 nm are due to carotenoids, the sorts of pigment responsible for the colors of carrots and tomatoes. These serve as accessory light-harvesting pigments and also have a protective function that will be discussed in Chap. 5.

Turning to the algae and green plants we find several new

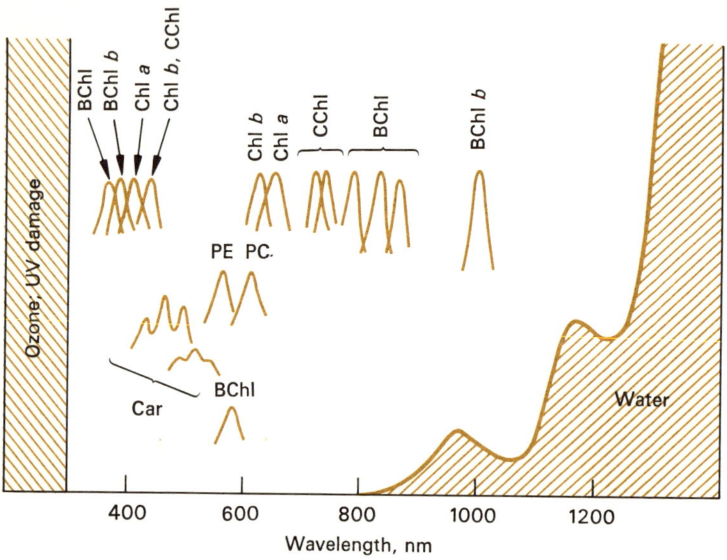

Fig. 1-14. The absorption peaks of light-harvesting pigments in various photosynthetic organisms fill the entire usable part of the spectrum (see Fig. 1-13). Chl, chlorophyll; BChl, bacteriochlorophyll; CChl, chlorobium chlorophyll; Car., carotenoid pigments; PE, phycoerythrin; PC, phycocyanin.

light-harvesting pigments that complete the utilization of every part of the visible spectrum, as indicated in Fig. 1-14. Especially important is the Chl *a* that is the major pigment of most higher plants and algae (see Fig. 1-4). In addition, particularly in land plants, we find Chl *b* in abundance, and carotenoids are universally present in photosynthetic organisms in nature.[1] Finally, there are the phycobilins, named after their structural similarity to the bile pigments that result from the degradation of hemoglobin. These include phycocyanin, absorbing at 630 nm and responsible for the blue tint of the blue-green algae, and phycoerythrin, absorbing at 570 nm and found in large amounts in the red algae. These pigments are open-chain tetrapyrroles (see the structure of a bilitriene

[1] A few laboratory pets, including a mutant strain of *R. spheroides* that yields the reaction centers shown in Fig. 1-11, lack carotenoids. These carotenoidless forms are killed by the combined action of light and oxygen; see Chap. 5.

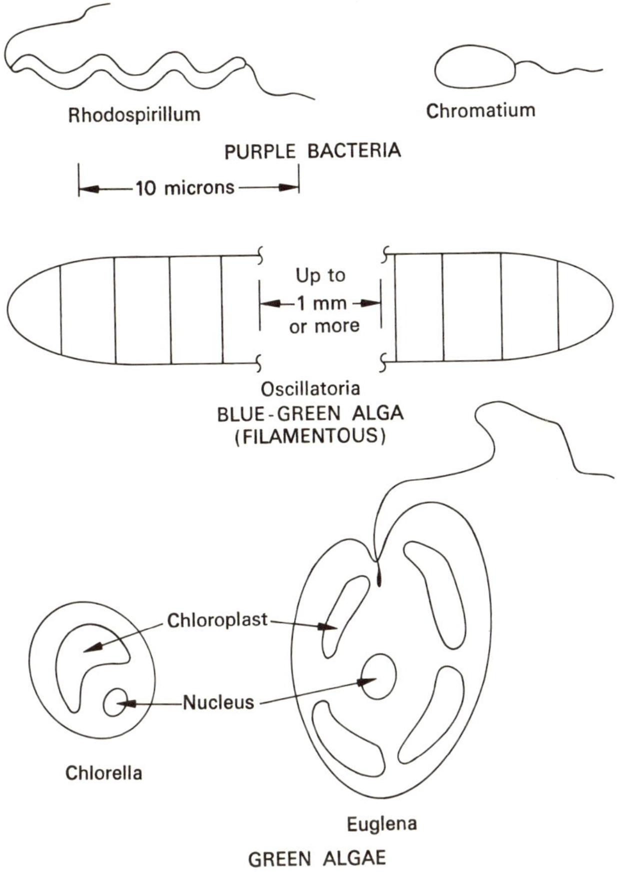

Fig. 1-15. Sketches of some microscopic photosynthetic organisms.

in Fig. 4-4); note their relationship to the chlorophylls, which have closed, or ring-shaped, tetrapyrrole structures.

Some algae and some photosynthetic bacteria are sketched in Fig. 1-15.

We have no convincing explanation for the intriguing correlation between the ability to evolve oxygen and the presence of Chl *a* in green plants and algae. The evolution of oxygen is a function of system II, but the Chl *a* serves both systems I and II. The P700 of the system I reaction center is probably Chl *a*. The question of which light-harvesting pigments serve which photosystems in green plants will be considered further in the next section.

THE COOPERATION OF TWO PHOTOCHEMICAL SYSTEMS IN PHOTOSYNTHESIS

1-8. Evidence for the Existence and Cooperation of Two Systems

If photosynthesis followed from a unique photochemical act, driven by quanta delivered from various light-harvesting pigments, the effectiveness of various wavelengths would depend on just two factors: the rate of light absorption and the efficiency with which the absorbed quanta are transferred to the reaction centers. An action spectrum for photosynthesis (see Sec. 3-13, Volume 1) should resemble the absorption spectrum of the tissue, with different pigments represented in proportion to the transfer efficiency. Any differences between different wavelengths should be solely quantitative; it should be possible to equalize the effects of light at two wavelengths by a suitable adjustment of the intensities.

By the early 1950s several observations on photosynthesis in algae had shown that the situation is more complicated. First, L. R. Blinks and others had seen that two widely different wavelengths, such as one in the green and one in the red, cannot be adjusted for strict photosynthetic equivalence. The intensities can usually be adjusted to give equal rates of photosynthesis (oxygen evolution and carbon dioxide assimilation) during sustained illumination, but the change from one wavelength to the other causes a transitory disturbance in rates no matter how the intensities have been adjusted. Furthermore, the transient in oxygen evolution has a different shape (when rate is plotted against time) than the one in carbon dioxide assimilation. These *chromatic transients* are sketched in Fig. 1-16*a*.

A second anomaly was the *red drop effect,* discovered in *Chlorella* by R. Emerson and C. M. Lewis, and later confirmed for many varieties of algae and green plants. Light of wavelength greater than about 680 nm is abnormally inefficient for photosynthesis, even though this light may be well within the absorption band of Chl *a*. This is seen when the action spectrum for photosynthetic oxygen evolution is compared with the absorption spectrum of the algae, as sketched in Fig. 1-16*b*. Of special interest was the later discovery in Emerson's laboratory that shorter-wave light (below 680 nm) could potentiate the effect of far-red light and bring its efficiency to a "normal" level (approximately equal to that of light below 680 nm). This *enhancement phenomenon* occurs even if the far-red and shorter-wave lights are presented as alternating flashes, separated in time by several seconds.

A third and wholly unexpected observation had to do with the transfer of excitation energy from phycobilins to Chl *a* and with the effectiveness for photosynthesis of light absorbed by these pigments. It appeared, to many observers including Blinks, Duysens, C. S. French, and F. Haxo, that the phycobilins in red and blue-green algae were unnaturally effective for photosynthesis. The action spectra for oxygen evolution showed that light absorbed by the phycobilins was used more efficiently than light absorbed by Chl *a* (Fig. 1-16*c*). In this connection the action spectrum for fluorescence of Chl *a* in the algae was particularly informative, since the fluorescence of a pigment is a direct expression of the presence of excitation energy in that pigment. The action spectrum for Chl *a* fluorescence paralleled the one for photosynthesis. This confirmed the expectation that light energy must be delivered to Chl *a* in order to be used for photosynthesis. However, it also established that the fluorescence of Chl *a* is sensitized more efficiently by phycobilins than by Chl *a* itself! To resolve this dilemma it had to be assumed that the algae contain Chl *a* in two states: one which is fluorescent, is effective in driving photosynthesis, and can receive energy from phycobilins, and the other which is nonfluorescent, inactive in photosynthesis, and inaccessible to the phycobilins. Direct excitation of Chl *a* goes into both types of the pigment, but excitation energy absorbed by phycobilins is transferred selectively to the active form of Chl *a*. The inactive form could then be identified as responsible for the red drop effect. This would suggest in turn that both forms are seen as one unresolved absorption band of Chl *a* in the algae, but that the inactive form

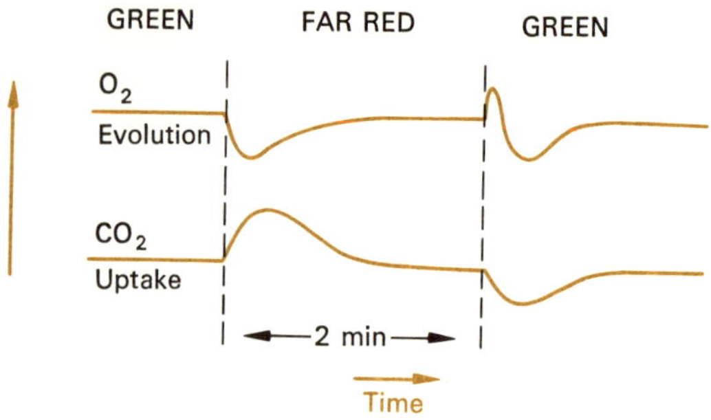

(a) Chromatic transients in algae

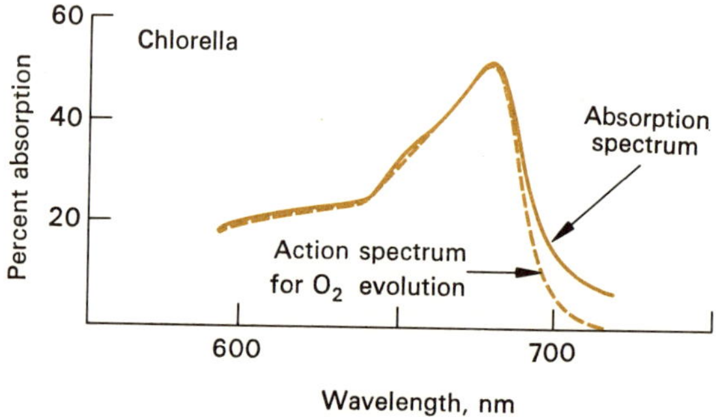

(b) The red-drop effect

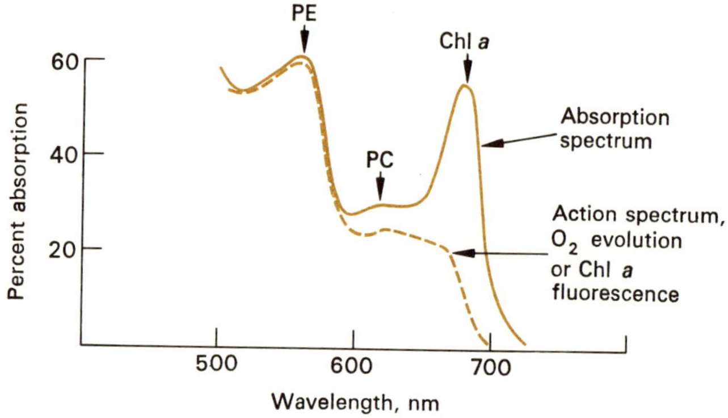

(c) Phycobilins are more effective than Chl *a* in sensitizing both photosynthesis and Chl *a* fluorescence.

contributes mainly to the long-wave part of the band, beyond about 680 nm. The enhancement phenomenon then shows that the "inactive" form of Chl *a* is not entirely so; it can be made active by the synergistic action of light below 680 nm.

This rather puzzling accumulation of facts showed that photosynthesis, at least in algae, involves more than one kind of light effect.

1-9. The Series Formulation, or *z* Scheme

The foregoing peculiarities were brought dramatically into focus by developments concerning cytochromes in plants. A cytochrome (Cyt) consists of an iron-containing tetrapyrrole molecule, called *heme,* bound covalently to a protein of low molecular weight (12,000 for mammalian Cyt *c*). The iron tetrapyrrole structure is similar to the magnesium tetrapyrrole of Chl. But the chlorophylls associate only weakly, through hydrophobic attractions rather than covalent bonds, with membrane proteins in the living tissue, whereas in the cytochromes the protein is a unique and characteristic part of the molecule. Cytochromes function as electron transport enzymes, cycling between their oxidized and reduced forms as they transmit electrons from one substance to another. Absorption spectra of mammalian Cyt *c* in its oxidized and reduced forms are shown in Fig. 3-11 of Volume 1. The designations *a, b,* and *c* were coined to distinguish different cytochromes that function in respiratory electron transport systems. The *c* types are *high-potential* (midpoint about +0.35 volt) forms that accept electrons from the lower-potential (about zero volt) *b* types and

Fig. 1-16. Three phenomena that suggest involvement of more than one light reaction in photosynthesis of green plants and algae. (*a*) Chromatic transients: *an abrupt change in the wavelength of light illuminating algae or leaves causes transient disturbances in the rates of oxygen evolution and carbon dioxide assimilation.* (*b*) The red drop effect: *light of wavelengths greater than about 680 nm, well within the absorption band of Chl* a, *is abnormally inefficient for photosynthesis.* (c) *The accessory pigments phycoerythrin (PE) and phycocyanin (PC) are often more effective than Chl* a *in sensitizing both photosynthesis and the fluorescence of Chl* a.

pass the electrons on to the *a* types that are directly involved with the reduction of O_2 in respiration. We now know that in bacterial photosynthesis, cytochromes of the *c* type give electrons directly to oxidized P870 following the primary photochemical act. Green plants contain a *c*-like cytochrome called Cyt *f*.

The 1950s saw the development of sensitive techniques for measuring small changes of light absorption in living tissues and the application of these techniques to the detection of subtle chemical changes including those attending photosynthesis. L. N. M. Duysens applied these methods to the study of light-induced reactions of both chlorophylls and cytochromes in photosynthetic bacteria and algae, and in 1961 he reported a striking observation concerning Cyt *f* in blue-green algae: Different colors of light have opposite effects on the redox state of this substance. In darkness the Cyt *f* drifts into its reduced form. Far-red light, such as 685 nm, then causes oxidation of the Cyt *f*, but if shorter-wave light is superimposed on the far red, the Cyt *f* is driven partly back to its reduced form. Thus far-red light, which had been associated with *inactive, nonfluorescent* Chl, causes oxidation of the cytochrome, whereas shorter-wave light, associated with *active, fluorescent* Chl, causes reduction.

R. Hill and F. Bendall had already had the insight, based on an accumulation of other observations about plant cytochromes, to propose a scheme like that of Fig. 1-9 (the *z* scheme, or series formulation) for photosynthesis in green plants and algae. Hill and Bendall suggested that cytochromes serve as electron carriers connecting the two photosystems, so that systems I and II should cause oxidation and reduction of these carriers, respectively. Duysens' observation could therefore be taken as confirming a major prediction of Hill and Bendall's model.

The red drop and enhancement effects could now be understood on the basis that far-red light drives system I alone but shorter-wave light drives both systems.

In the past decade the skeletal *z* scheme of Fig. 1-9 has been given the substance of detail, through a variety of techniques for studying the two systems separately and together. The major techniques have been:

1. *Chemical isolation of the two systems.* As one example, the herbicide dichlorophenyldimethylurea (DCMU) blocks the flow of electrons from system II into the path connecting systems II and I. This inhibitor can be added, and the usual flow of electrons to system I from system II can then be replaced by electrons from a reducing agent such as reduced indophenol.[1] The reduction of $NADP^+$ to NADPH, a function of system I, can then be studied without the involvement of system II. The metabolism of Gaffron's hydrogen-adapted algae (Sec. 1-2), where electrons from H_2 replace those from system II, can be taken as a special case of this kind. As a second example, the connection between systems I and II can be weakened, in suspensions of chloroplasts[2] derived from leaves, by washing away soluble components. Then an electron acceptor such as ferricyanide can be added, to intercept the flow of electrons from system II and prevent them from reaching system I. The result is a Hill reaction (see Sec. 1-2) involving just system II, as distinguished from one in which electron flow through both systems has been kept intact and the artificial acceptor receives electrons from the top of system I.

2. *Genetic dissection.* Mutant algae have been developed, especially by R. P. Levine and by N. I. Bishop, that lack certain electron transport components such as cytochromes or even P700 (the latter mutants are analogous to Sistrom's P870-less mutants of photosynthetic bacteria). These deficiencies cause the operation of one photosystem, or the continuity of associated electron transport chains, to be lost. By studying such mutants one can piece together the normal sequence of components. For example, a mutant of the alga *Chlamydomonas* lacks plastocyanin, a copper-

[1] An effective system introduced by L. P. Vernon is a mixture of ascorbic acid and 2,6-dichlorophenol indophenol, the former as ultimate source of the electrons and the latter to couple the flow of electrons into a site near the "bottom" of system I.

[2] Chloroplasts are the subcellular bodies that contain the photosynthetic apparatus in higher plants, and in most algae.

containing protein that functions in electron transport between systems I and II. Cyt *f* in the mutant cannot be oxidized by far-red light, but P700, the reaction center Chl of system I, is oxidized. This places plastocyanin between Cyt *f* and P700. Operationally related to genetic dissection is the technique of washing out soluble components, noting loss of activity, and adding them back to restore activity.

3. *Physical separation.* Following the lead of M. B. Allen and of N. G. Boardman and J. M. Anderson, many investigators have succeeded in fractionating plant chloroplasts into classes of smaller particles that are enriched for components and reactions of system I or II. The chief method is to attack suspensions of chloroplasts with detergents and then to effect a separation of "heavy" and "light" particles by centrifugation through an aqueous sucrose density gradient. This approach is important, because it shows that the two systems exist as separate physical entities.

4. *Kinetic absorption spectrometry.* The dynamics of optical absorption changes following a flash of light show the conversions of various components (P700, cytochromes, quinones, etc.) between their oxidized and reduced forms. Kinetic analyses can then reveal the relationship of these components to each other and to systems I and II. This approach has been pursued especially by H. T. Witt, as well as by B. Kok, B. Chance, Duysens, and others.

The foregoing analytical approaches have been continually interlaced with the following questions:

1. Can the sequence of interactions between components be decided unequivocally?

2. Are the reactions of quantitative significance? This is usually settled by measuring the quantum efficiency. For example the stoichiometry between far-red light quanta absorbed by system I, electrons delivered through P700, and molecules of $NADP^+$ reduced shows that $NADP^+$ reduction coupled with P700 oxidation can be a major activity of system I.

3. What light-harvesting pigments serve the two photosystems? This is decided by measuring action spectra for the various partial reactions associated with systems I and II, and also for

phenomena such as the enhancing action of shorter-wave light in relation to the red drop. These action spectra have confirmed that system I is driven mainly by a far-red form of Chl *a,* with an absorption peak at 683 nm, which accounts for most of the absorption beyond 680 nm (historically this is the nonfluorescent, inactive form). Phycobilins are especially effective in driving system II by delivering energy to a component of Chl *a* (fluorescent; active; absorption peak at 672 nm) associated with that system. Chl *b* is also associated with system II.

Some principal results of investigations along these lines during the 1960s are summarized in Fig. 1-17, an expanded version of Fig. 1-9. The main embellishment has been the explicit formulation of electron flow from the top of system I to a point between systems I and II, establishing a basis for cyclic operation of system I coupled to phosphorylation. The specific number and location of coupling sites, as well as the mechanism of coupling, remain uncertain.

The association of certain light-harvesting pigments with each system is not entirely rigid. Energy absorbed by the pigments generally associated with system II can "spill over," to some extent, to system I. Variation of the spillover provides the basis for a regulatory system that will be discussed later.

One can concoct all kinds of variations and embellishments on the scheme of Fig. 1-17. As shown it represents a reasonable consensus as of 1970.

1-10. System II: Oxygen Evolution, Electron Transport, and Chlorophyll Fluorescence

The mechanism for oxygen evolution in photosynthesis is represented in Fig. 1-17 by the arrows leading from H_2O to Chl a_{II} at the hypothetical reaction center of system II. One could say alternatively that excited Chl a_{II} delivers electrons to Q, and $\oplus$'s (oxidizing equivalents) toward H_2O. Four $\oplus$'s must then be assembled in order to take four electrons away from $2H_2O$ and yield one O_2, together with four H^+ ions:

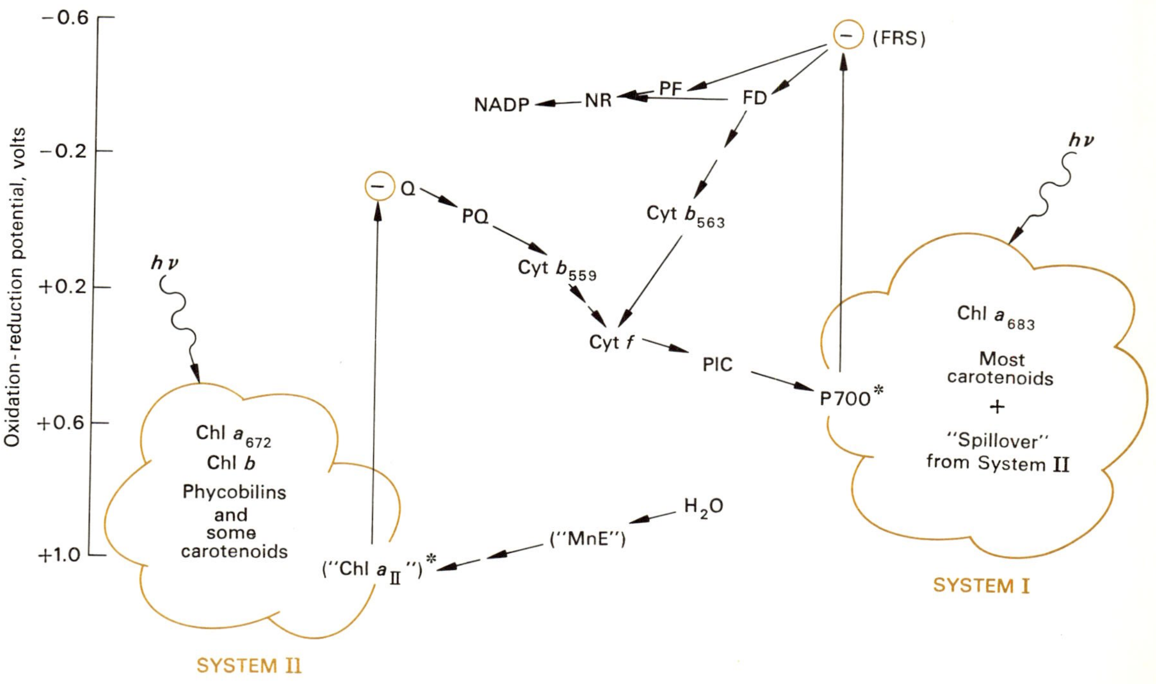

Oxidation-reduction potential, volts
−0.6
−0.2
+0.2
+0.6
+1.0
(FRS)
NADP
NR
PF
FD
Q
PQ
Cyt b_{563}
Cyt b_{559}
Cyt f
PIC
P700*
$h\nu$
Chl a_{683}
Most carotenoids
+
"Spillover" from System II
SYSTEM I
$h\nu$
Chl a_{672}
Chl b
Phycobilins and some carotenoids
("Chl a_{II}")*
("MnE")
H_2O
SYSTEM II

Fig. 1-17. *A more detailed picture (see Figs. 1-8 and 1-9) of the cooperation of two photosystems in the photosynthesis of green plants and algae. Electrons are driven through the system as indicated by the arrows; a sequence of two arrows shows the likelihood of one or more unknown components between those shown. The vertical position of each component indicates its approximate reducing potential. The long upward arrows terminating at ⊖ denote the photochemical steps in which electrons are promoted to higher energies by excited Chl* a_{II} *or P700 (excitation is symbolized by the asterisks). Parentheses around a component mean that its identification is presumptive. Sites where electron transport is coupled to phosphorylation are uncertain and are not indicated, but there is at least one in the electron flow cycle driven by system I. The principal light-harvesting pigments for each system are shown. Abbreviations (other than those defined in the text) are as follows:*

MnE = *hypothetical enzyme containing manganese*
Chl a_{II} = *presumed "reaction center chlorophyll" that engages in the photochemistry of system II*
Z = *hypothetical donor of electrons to oxidized Chl* a_{II}
Q = *primary electron acceptor in the photochemistry of system II; may be a quinone*
PQ = *plastoquinone*
PlC = *plastocyanin*
NR = *NADP reductase*
PF = *phytoflavin*
FD = *ferredoxin*
FRS = *ferredoxin-reducing substance, possibly a pteridine-protein complex*

Further description of these compounds, beyond that in the text, is outside the scope of this book.

$$2H_2O \longrightarrow 4H^+ + 4e^- + O_2 \qquad (1\text{-}20)$$

Elucidation of this process has been a peculiarly refractory problem, perhaps because the components are tightly bound to membranes in the chloroplast. This could make it difficult to isolate the components in soluble and yet functional form; there is evidence that membrane fragments large enough to hold about 1,000 Chl molecules are needed for the preservation of oxygen-evolving activity. Also the close binding could make the turnover of each component so rapid as to prevent detection of transitory intermediates (the delay between a brief flash of light and the evolution of O_2 can be less than a millisecond). The fact is that we have not identified with certainty any substance on the pathway from H_2O to Q. The characterizations that come through the analysis of optical absorption changes, which have been so informative for system I and with photosynthetic bacteria, have told us little about system II.

Lacking a detailed knowledge of the components of system II, we have had to extract information from interrelated peripheral phenomena: the flow of electrons out of the system, the evolution of O_2, and the fluorescence and other luminescences of Chl *a*. The information is interpreted in terms of a simple model:

$$\begin{array}{ccc} H_2O \quad \cdots & Z\cdot Chl\cdot Q & \cdots A \\ & \downarrow h\nu & \\ & Z\cdot Chl^*\cdot Q & \\ & \downarrow & \\ & Z^+\cdot Chl\cdot Q^- & \\ & \downarrow & \\ H_2O \quad \cdot\cdot(+)\cdot\cdot & Z\cdot Chl\cdot Q & \cdots A^- \end{array} \qquad (1\text{-}21)$$

In this picture Chl represents both the light-harvesting and the photochemical function of Chl *a* in system II. The fictitious MnE (Fig. 1-17) has been omitted; Z is the nearest neighbor of Chl on the positive side of the reaction center, and Q on the negative side. "A" represents the totality of natural and artificial electron acceptors such as the native plastoquinone (PQ) and any oxidants such as ferricyanide that might be introduced. Four ⊕'s must cooperate to give one O_2 from water.

The clearest and simplest correlation is between the fluorescence of Chl and the flow of electrons from Q^- to A. When electrons are prevented from leaving Q^-, the intensity of flourescence rises. Apparently the photochemical utilization of light quanta depends on Q being in its oxidized form. As soon as it accepts an electron and becomes Q^-, the photochemical act is blocked, and more of the quanta are returned to the surroundings as fluorescence (see Sec. 2-15, Volume 1). One way to show this effect is to add a strong reducing agent to a suspension of chloroplasts, so as to convert the hypothetical Q to Q^-, and observe that the fluorescence becomes stronger. One might expect a similar effect if the hypothetical Z is converted to its oxidized form. However, the addition of oxidants only depresses the fluorescence, probably by acting on Q^-. Either Z is difficult to oxidize chemically, or else the state of Z does not affect the primary photochemistry. This suggests that the only photochemical act that competes with fluorescence is

$$\text{Chl}^* \cdot \text{Q} \longrightarrow \text{Chl}^+ \cdot \text{Q}^-$$

and that the transfer of ⊕'s from Chl^+ to Z is a "dark" sequel to the photochemical act.

The change in the intensity of fluorescence in a suspension of spinach chloroplasts, during illumination that drives the photochemistry of system II, is shown in Fig. 1-18. During the first few seconds the fluorescence rises as the Q's are converted to Q^-'s. If ferricyanide is then injected into the sample, the fluorescence falls again, because this reagent takes electrons from Q^-. After all the ferricyanide has been reduced in this Hill reaction, the fluorescence rises once more. It can be shown that the amount of ferricyanide is proportional to the area *above* the fluorescence curve, in the region of the ferricyanide-induced dip. The area above the "initial-rise" part of the curve then measures the amount of native electron acceptor that can take electrons from Q^-. Finally, the

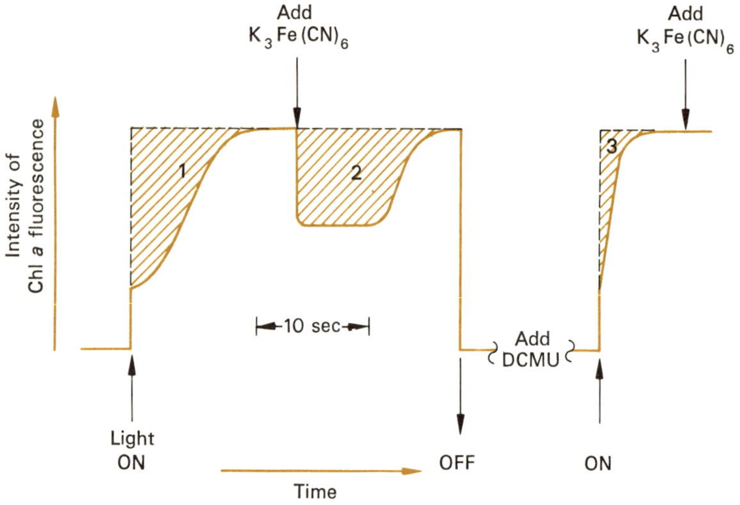

Fig. 1-18. Fluorescence of Chl a *in spinach chloroplasts, during a program of illumination and injection of ferricyanide. See the text for discussion. The shaded areas above the curves are proportional to the amounts of electron acceptors for system II: area 1, native acceptors; area 2, added ferricyanide; area 3, acceptor Q (DCMU appears to prevent the flow of electrons from Q to the other acceptors). The time scale is somewhat arbitrary; it depends on the rate of electron flow and hence on the light intensity.*

electron acceptors (both native and artificial) can be cut off from Q by adding DCMU, which appears to block electron transfer just beyond Q. The area above the initial rise in the fluorescence curve is then much smaller, reflecting only the amount of Q itself. Quantitative studies of this kind, by S. Malkin and by P. Joliot, indicate that there are about 300 to 500 molecules of Chl *a* for every Q. This tends to define the size of the "system II photosynthetic unit," making it about the same as that of system I (there are about 400 Chl *a*'s for every molecule of P700). But the amount of native A, which is probably plastoquinone for the most part, is about tenfold greater than that of Q in terms of electron equivalents. Also there are two molecules of Cyt b_{559}, one of Cyt *f*, and one of plastocyanin for each P700. We should therefore imagine that the pathways of electron flow converge from many A's (identified as plastoquinone in Fig. 1-17) to each P700.

Photosynthetic oxygen evolution in chloroplasts can be inhibited by certain washing procedures, by mild heat treatment, or by ultraviolet irradiation. These treatments impair the flow of electrons from H_2O to reaction centers, but this source of electrons can then be replaced by an artificial electron donor such as aminophenol, $H_2N(C_6H_4)OH$. System II can then function without O_2 evolution but with the ability to deliver electrons to system I:

Electrons from H_2O —⟩⟨→
$(+) \cdots Z \cdot Chl \cdot Q \cdots A^- \rightarrow$ electrons to system I
Electrons from aminophenol —↑

(1-22)

This discovery, made by T. Yamashita and W. L. Butler, will hopefully lead to more direct chemical investigation of the components between H_2O and the reaction centers of system II.

Another key to understanding the chemistry of O_2 evolution is the *activation effect,* discovered by F. L. Allen and J. Franck and exploited most effectively by P. Joliot. Consider a suspension of algae or chloroplasts exposed to a prolonged sequence of brief, strong flashes of light, say a flash of duration $<10^{-5}$ sec given every 2 sec. The duration is kept short enough to allow just one photochemical cycle at each reaction center. Under these conditions the "normal" yield of photosynthetic O_2 evolution is about 1 molecule per flash for every 3,000 Chl molecules present. Now if the suspension is kept dark for several minutes and then the flashing is resumed, a curious pattern can be seen in the responses to the first several flashes. The very first flash after a dark period elicits no O_2 evolution, no matter how strong the flash. The second flash, given within a few seconds after the first, again yields almost no O_2. However, these flashes prepare the system so that a third flash, given a few seconds after the first two, gives a supernormal yield: about 2 to 3 O_2 molecules per 3,000 Chl molecules. The yield of the fourth flash is near normal (1 O_2 per 3,000 Chl). Flashes 5 through 8 show a damped, or less striking, version of the same pattern as flashes 1 through 4: subnormal for 5 and 6, supernormal for 7, and near normal for 8. The damping effect continues, so that after about 20 flashes the variations are washed out and each flash elicits a "normal" yield of O_2. The variations can be restored by allowing the material to rest in the dark for about 2 min.

This curious pattern of oscillations in the flash yield can be understood if four $\oplus$'s must be put together before one molecule of O_2 can be released. After a time in darkness there are few $\oplus$'s remaining in the system. The first flash makes one at each reaction center, the second another, and so forth. When enough $\oplus$'s are present, they are utilized en masse, discharging the system in a burst of supernormal activity. Then the accumulation of $\oplus$'s must begin anew. We have no knowledge of how many chemically distinct kinds of $\oplus$'s participate in this activity, but the pattern of variations fits a simple model: After dark adaptation most of the reaction centers have just one $\oplus$. After one flash these reaction centers have two $\oplus$'s apiece, and after two flashes they have three $\oplus$'s. Each of these reaction centers is then poised so that a third flash gives four $\oplus$'s, which are discharged in the liberation of one O_2 molecule (supernormal effect). Some of the reaction centers do not attain this condition until the fourth flash, but most of them are discharged and must start over, building up the condition of four $\oplus$'s again with the seventh flash.

A conjunction of flash yield analysis with electron donation in the manner of Yamashita and Butler may tell us more.

All photosynthetic tissues emit delayed fluorescence from Chl or BChl. This emission, which has the spectrum of the prompt fluorescence but a much greater lifetime, represents a reversal of early physical or chemical events so as to repopulate the singlet excited state of the pigment (see Sec. 2-11, Volume 1). Delayed fluorescence in *Chlorella* cells was discovered accidentally by B. L. Strehler and W. Arnold during an effort, in 1951, to detect photosynthetic phosphorylation with the firefly luminescence technique.

An extract of firefly tails gives an easily measured glow in response to the addition of a minute amount (such as 10^{-10} mole) of ATP. Strehler and Arnold found that *Chlorella* cells, previously illuminated, gave luminescence even when the firefly extract was omitted from the test. They soon established that the emission was from the same excited state of Chl *a* that gives fluorescence. Numerous studies since that time have established that in green plants and algae the delayed fluorescence is almost exclusively a property of system II, that it is related to the functioning of the reaction centers, and that it probably involves recombination of some of the $\oplus$'s and $\ominus$'s that are generated at the reaction centers. Thus the delayed fluorescence might be developed as another experimental approach to the presence of the $\oplus$'s that lead to O_2 evolution.

One peculiarity of the delayed fluorescence is that its intensity is reduced strongly by substances that render biological membranes leaky toward ions. We shall return to this problem in connection with membrane function in the photosynthetic bacteria.

1-11. The Relationship between Photosynthetic Bacteria and Green Plants

When the *z* scheme was announced, it seemed natural to regard the bacterial photosynthetic system as analogous to green plant system I. The analogy became especially compelling when Bishop and Gaffron showed that the "bacterial" photosynthesis performed by hydrogen-adapted algae utilizes system I exclusively. Both the bacteria and system I can remove electrons from substrates such as H_2, and both can drive electrons through a cycle coupled with phosphorylation. Both have reaction centers based on a specialized Chl or BChl (P700 or P870) that undergoes reversible oxidation and reduction at a midpoint potential of about +0.45 volt.[1] However, the potentials generated on the reducing side appear to be quite different.

[1] P700 and P870 can be oxidized and reduced chemically. For either pigment the concentrations of oxidized and reduced forms are about equal when the oxidation-reduction potential has been adjusted to a value near +0.45 volt by means of a mixture of ferri- and ferrocyanide.

The reducing potential can be estimated by illuminating the photosynthetic material in the presence of a reducible substance of known midpoint potential. Reduction of a high proportion of the substance then indicates that its midpoint potential is less negative than that generated by the photochemical system. Illuminated suspensions of chloroplasts or of subchloroplast particles enriched for system I cause extensive reduction of viologen dyes having midpoint potentials of about —0.6 volt. In contrast, illuminated bacterial chromatophores appear unable to reduce substances of midpoint potential lower than about —0.1 volt.

Further information about the reducing potential of the primary electron acceptor can be gained by studying the fluorescence of BChl in chromatophores or of P870 in reaction center preparations. The rationale is like that for system II: when the acceptor becomes reduced, the photochemical act is blocked, and the fluorescence rises. The rise in fluorescence caused by the addition of reducing agents again suggests that the primary acceptor is reduced at a potential of about —0.1 volt.

These experiments, which indicate that the primary acceptor has a stable reduced state at —0.1 volt, do not rule out the possibility of a (less stable) reduced state of higher energy, say —0.6 volt, that can be formed photochemically. Such a state could escape detection in preparations of chromatophores and reaction centers, because of a rapid decay to the lower (—0.1 volt) state. But in the living cell, with close proximity of secondary electron acceptors, the electron might be transferred and stabilized at the higher energy. Thus the experiments showing a midpoint potential of —0.1 volt do not necessarily reveal the full capability of the system.

In cells of the photosynthetic bacterium *Chromatium* there are two cytochromes that undergo light-induced oxidation, through reactions with oxidized P870. These are Cyt 553 and 555, named after the location of the *alpha peak* in the absorption spectrum. In weak light the oxidation of Cyt 553 takes precedence over that of Cyt 555, but the former can be taken out of the picture because it is susceptible to oxidation by O_2. The light-induced oxidations of the two cytochromes can therefore be studied independently:

Cyt 553 in weak light in the absence of O_2 and Cyt 555 in the presence of O_2 or as an incremental effect in stronger light. S. Morita has shown that the action spectra for the oxidations of Cyt 553 and 555 are different, reflecting participation by different proportions of the three main light-harvesting components of BChl in *Chromatium*. Meanwhile, accessory biochemical experiments had indicated that Cyt 555 is involved with cyclic electron flow coupled to phosphorylation, whereas Cyt 553 participates in a noncyclic pathway that mediates oxidation of substrate. It could therefore be proposed that the bacteria, like the green plants, have two photochemical systems: one for cyclic and one for noncyclic electron transport. Similar observations and interpretations have been made by C. Sybesma, working with *Rhodospirillum rubrum*.

At present the evidence for two systems in the bacteria is still tenuous and beset by internal contradictions. So far there is no evidence for a photochemical act with characteristic oxidizing and reducing potentials very much different from +0.45 and —0.1 volt, the approximate values exhibited by the components of the reaction center preparations described earlier. The "two photosystems" in bacteria might be nothing more than regions in the tissue that differ in the relative abundances of various secondary electron transport components. The different action spectra could then reflect variations in the physical state of the BChl in these regions.

1-12. The Dynamics of Photochemistry in the Bacteria

The fluorescence of Chl or BChl can give information about the rates of events leading to photochemistry, because the yield (or quantum efficiency) of fluorescence is reduced in proportion to the rates of the processes that compete with fluorescence (see Secs. 2-15, 2-16, and 3-7 in Volume 1). In chromatophores the yield of BChl fluorescence is about 5 percent. If the yield were 100 percent, the lifetime of excitation would be [from Eq. (3-17), Volume 1] about 2×10^{-8} sec. The yield of 5 percent then corresponds to a lifetime twentyfold shorter, or about 10^{-9} sec. This is the average time needed for the transfer of excitation quanta from light-harvesting BChl to the reaction centers where they are used. A value of 10^{-9} to 2×10^{-9} sec has been verified by direct measurements of the decay of BChl fluorescence in photosynthetic bacteria.

In reaction center preparations the fluorescence from P870 competes with the photochemical act itself, rather than with a transfer of energy to the sites of photochemistry. The yield of this fluorescence is about 0.05 percent, corresponding to a lifetime of about 10^{-11} sec. This is presumably the time needed for the photochemical electron transfer.

The earliest events following the photochemical act in the bacteria are the transfer of the electron from the primary acceptor to a pool of secondary acceptors and the transfer of an electron from a cytochrome to P870. The kinetics of these events have been studied by applying brief flashes of light (shorter than 10^{-7} sec, from a laser) to chromatophores or cells and recording the subsequent changes of absorption and fluorescence that indicate the states of cytochromes, P870, and primary acceptor. Such investigations by B. Chance, W. W. Parson, and others show that these electron transfers require times ranging from about 10^{-6} to 10^{-3} sec. At low temperatures these events are sometimes slowed to the point that they no longer supervene over a direct reversal of the primary photochemistry (a return of electrons directly from primary acceptors to oxidized P870). The direct reversal can be seen, at temperatures approaching 1°K, to require about 20 msec in *Rhodopseudomonas spheroides.* At room temperature this potentially wasteful process is circumvented by the much faster movements of electrons from cytochromes to oxidized P870 and from primary acceptor to secondary pool.

STRUCTURE AND FUNCTION: MEMBRANES AND REGULATORY SYSTEMS

1-13. The Structures of Photosynthetic Tissues

When thin sections of photosynthetic bacteria are examined in the electron microscope, it can be seen that the cell membrane, just inside the cell wall, is extensively convoluted. Some of the invaginations often seem to have become pinched off, forming round membrane-bound objects within the ceil (see Fig. 1-19*a*). It would be tempting to equate these spherules with the chromatophores that are obtained from broken cells, but the chromatophore population undoubtedly includes fragments of the originally intact membrane. The fragments, usually about 300 to 1500 Å in diameter, appear to have closed upon themselves so as to form an inner space entirely enclosed by a membrane, because they exhibit controlled uptake and release of ions during illumination. The inner phase of the chromatophore might be topologically inverted in relation to the inner phase of the cell; it certainly would be in the case of a pinched-off invagination, where a parcel of the exterior has become enclosed in the spherule.

Sometimes the membrane is seen to form multiple stacks of flattened bags, in a lamellar structure as sketched in Fig. 1-19*b*. The intracellular membrane-bound structures have the general name of *thylakoids* (Greek for baglike).

The membrane appears double, with a thickness of about 100 Å. It undoubtedly holds the light-harvesting pigments and the reaction centers, because these components are carried with the subcellular fraction that can be identified as membranous. One can make detailed models of this double membrane, fitting together the known proportions of protein, lipids, pigments, etc., to form a structure of the proper size and general properties, but this is merely an exercise in geometrical chemistry because we lack firm information about the disposition of these components.

The photosynthetic membrane in cells of green plants and most algae is organized inside one or more chloroplasts, subcellular organelles that are usually several microns in linear dimension. The lamellar thylakoids extend throughout the chloroplast, and in some chloroplasts there are local regions of greater density called *grana* (Fig. 1-19*c*). The grana are usually about one micron in extent.

As with photosynthetic bacteria, the chloroplast structure shows an emphasis on membrane-enclosed spaces. The usual recipes for making suspensions of chloroplasts (disruption of the leaf, followed by differential centrifugation) yields many fragments which are smaller than chloroplasts, but which preserve the integrity of a membrane-bound inner phase.

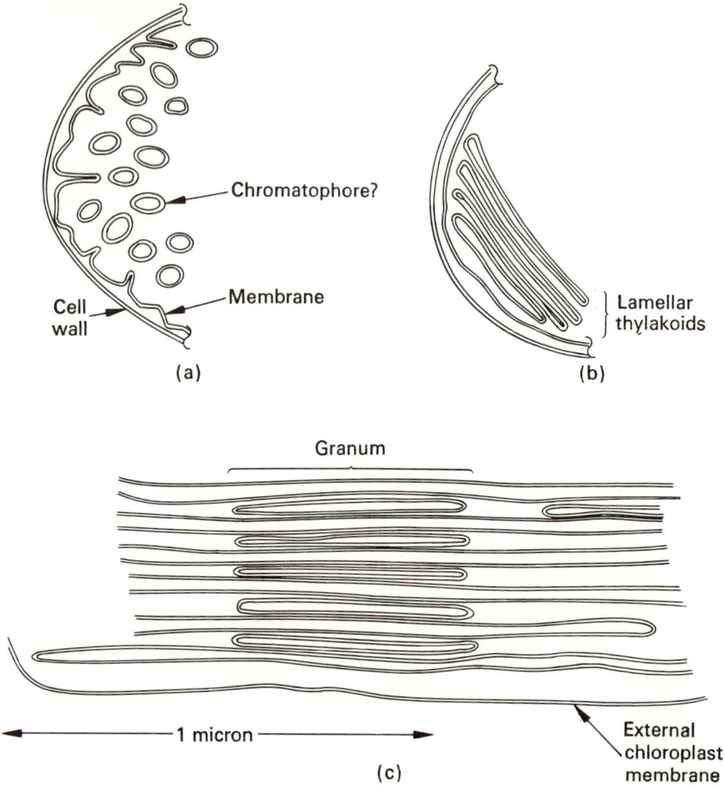

Fig. 1-19. Sketches of photosynthetic membrane structures. (a) The cell membrane in a photosynthetic bacterium is convoluted, and some of the invaginations appear to pinch off, forming round inclusions. (b) Sometimes the membrane in a photosynthetic bacterium becomes differentiated to form an extended structure of lamellar thylakoids (flattened bags). (c) The chloroplasts of green plants and algae contain lamellar thylakoids; in some species these are arranged so as to form denser regions called grana.

Electron microscopy shows that the surface of the thylakoid membrane in green plants is bumpy, both as a reflection of internal structures and because small particles are attached to the outside. The internal bumps might correspond to photosynthetic units (system I, or II, or both?), and the external ones might be multienzyme systems involved in phosphorylation and/or carbon dioxide fixation. These assignments are not yet certain.

1-14. The Coupling between Electron Transport and Phosphorylation

Photosynthetic electron transport is attended by the translocation of ions across the thylakoid membranes, most notably the movement of protons into the inner space. This pumping of ions leads, through osmosis, to the movement of water and hence to swelling or shrinking of the thylakoids. The ion flow is also attended by changes in the electric potential across the membrane. Factors that nullify either the ion gradients or the membrane potential, or both, usually cause the system to lose its ability to make ATP. Apparently the coupling between energy-yielding electron transport and energy-storing phosphorylation involves a high-energy intermediate, and this intermediate (whether it be a substance or a state of the entire thylakoid) is associated with the ion gradients and the potential across the membrane.

The earliest attempts to understand the coupling of electron transport to phosphorylation were based entirely on chemical considerations. The formation of ATP from ADP and inorganic phosphate involves a dehydration: the creation of an anhydride bond. A link between the energy of electron transport and that of phosphorylation might therefore be a substance whose chemistry involves both oxidoreduction and hydration-dehydration. This is the essence of chemical coupling hypotheses advanced by F. Lipmann and developed by E. C. Slater, A. L. Lehninger, B. Chance, and others. A typical formulation, to link a reaction of the type

with

$$ADP + P_i \longrightarrow ATP + H_2O$$

invokes an intermediate I that can form an anhydride bond with Y, and a substance X that can transfer the bond energy:

$YH_2 + I \longrightarrow YH_2 - I$ (anhydride bond formation)

$YH_2 - I + Z \longrightarrow Y \sim I + ZH_2$ (the anhydride bond gains in energy as YH_2 is oxidized to Y)

$Y \sim I + X \longrightarrow Y + X \sim I$ (the bond is transferred to X)

and finally

$$X \sim I \curvearrowright XH \cdots HOI \rightarrow X + I + H_2O$$
$$ADP + P_i \curvearrowright ATP$$

($X \sim I$ takes H_2O from ADP + P_i, transferring the anhydride bond energy to ATP)

The difficulty with schemes of this kind is that no one has been able to identify (much less isolate) the essential chemical intermediates. Also this kind of scheme does not explain why phosphorylation is associated with membranous structures (membranes of mitochondria, of bacteria, or of chloroplasts) that exhibit a controlled translocation of ions and water during the phosphorylating activity.

To meet these problems P. Mitchell proposed that the coupling is not by chemical substances but by electrochemical states: a difference in the concentration of H^+ on two sides of a membrane and the electric potential across the membrane. A generalized representation of this idea is shown in Fig. 1-20. It involves two postulates. The first is that electron transport is obligately and stoichiometrically related to the movement of H^+ across the membrane that separates an inner from an outer phase. This is indicated at the top of Fig. 1-20, and a more concrete possibility is shown in Fig. 1-21. The second postulate is that the ATP-forming enzyme system (ATPase) is located in a specially polarized part of the membrane: when water is released in the condensation of ADP

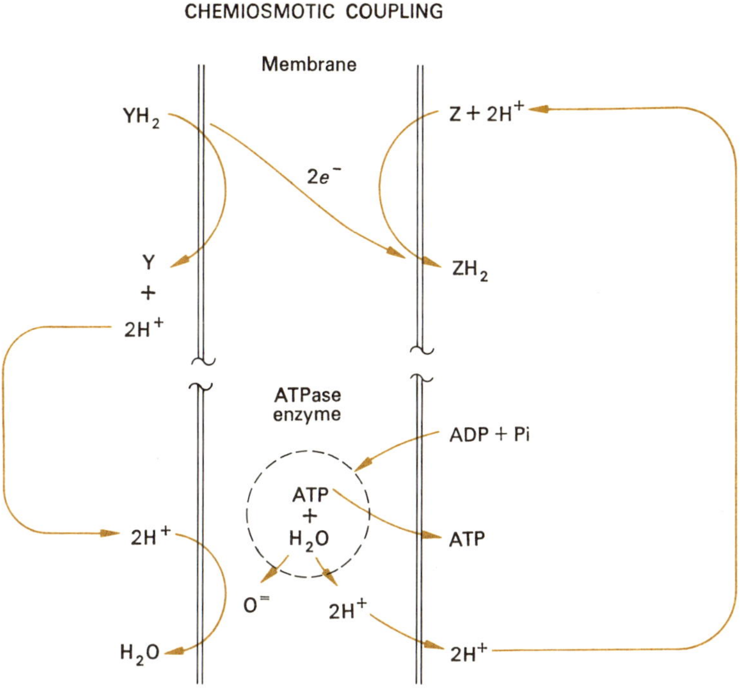

Fig. 1-20. A representation of P. Mitchell's hypothesis for the coupling of electron transport to the formation of ATP (see the text). The energies of oxidoreduction and phosphorylation are linked through the common electrochemical parameter of H^+ *concentration. The mechanism requires a structure bounded by a membrane; in this figure either side of the membrane could correspond to the inside of the structure. The species* $O^=$ *would probably be carried by chemical groups associated with the ATPase enzyme system.*

and P_i, the H^+ can come out only on one side and OH^- or $O^=$ on the other. The choice of $O^=$ is made in order to satisfy an observed stoichiometry of two H^+ translocated for each ATP formed. In this picture the transport of electrons and the formation of ATP are linked thermodynamically by the H^+ concentration in the inner phase of the structure. The inner phase is arbitrary in Fig. 1-20. If we take it to be on the right, we can say that electron transport propels the ATP-forming reaction, through the mass-

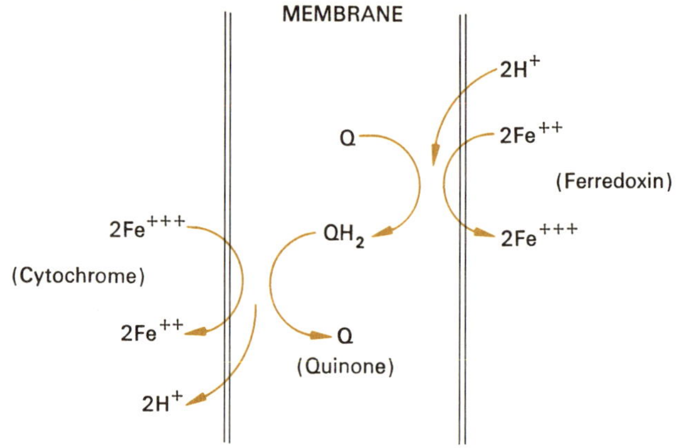

Fig. 1-21. A more specific model for proton translocation coupled to electron transport across a membrane (compare with the upper part of Fig. 1-20). The fat-soluble quinone could move in the lipid environment of the membrane.

action principle, by drawing away H^+. An electric potential across the membrane will arise if different ions (positive and negative) are pumped or can diffuse at different rates. This potential enters into the thermodynamics of coupling, by facilitating or opposing the transport of charge across the membrane.

The choice between the chemical and the chemiosmotic mechanisms is still in contention. A basic question that has not been resolved is whether the observed pumping of H^+ across a membrane is a necessary state in the coupling or is only a byproduct of chemical activity.

Attention was drawn to the chemiosmotic hypothesis especially by a striking discovery made by A. T. Jagendorf, G. Hind, and E. Uribe: A suspension of chloroplasts can make ATP in response to an artificial pH gradient. The gradient was created by first soaking the chloroplasts in a medium of pH 4 and then transferring them abruptly to a medium of pH 8. The result was a temporary gradient of H^+ concentration across the thylakoid membrane, with $[H^+]$ greater on the inside. This treatment mimicked the result of the physiological light-induced pumping of H^+ into the thylakoids. If ADP and P_i were added when the gradient was created, ATP was formed. The quantitative aspect of the experiment was remarkable: the amount of ATP formed by this treatment was equivalent to the product of about 100 cycles of the photochemical machinery. It appeared that the energetic state envisioned by Mitchell could indeed bring about the expected formation of ATP.

This experiment can be explained in terms of a chemical coupling hypothesis by supposing that the $[H^+]$ gradient shifts the equilibria of reactions that involve the coupling substances. However, the explanation based on Mitchell's hypothesis is especially direct and satisfying.

1-15. Uncouplers: Ion Transport and Membrane Potential

Under physiological conditions the rate of electron transport is governed by the rate at which ATP can be formed. If the entire sequence of events is blocked at any one point, the earlier steps in the sequence are inhibited, suggesting that a mass-action principle is operating. Thus if the supply of ADP is depleted or if the ATPase enzyme is inhibited, the failure of ATP formation is reflected in a slowing of electron transport. On the other hand a variety of substances, called *uncouplers,* seem to break the connection, so that the earlier steps (electron transport) are no longer constrained by the rates of the later ones (ATP formation). We can understand the action of an uncoupler if we imagine that it destroys or consumes the hypothetical "coupling intermediate," preventing the formation of ATP while allowing electron transport to go untrammeled. In terms of Mitchell's chemiosmotic hypothesis, the coupling intermediate is a combination of H^+ and other ion gradients across the membrane. An uncoupler should then be any agent that dissipates these gradients.

There are, indeed, many substances which interfere with ion gradients and electric potentials across thylakoid membranes and which act as uncouplers. Among these antibiotic substances are valinomycin,[1] which makes the membranes leaky to K^+ and NH_4^+; gramicidin, which induces permeability toward many univalent cations; and nigericin, which exchanges K^+ for H^+ across the membrane.

If valinomycin is added to a suspension of chloroplast fragments or chromatophores, any membrane potential due to the pumping in of H^+ ions can be lost because of a compensating outflow of K^+. In fact, the membrane potential in the presence of valinomycin is indicated by the ratio of internal to external K^+ concentration.[2] Gramicidin, like valinomycin, can cause the collapse of a photosynthetically induced membrane potential. Nigericin can allow the dissipation of a proton gradient without loss of the membrane potential, by bringing K^+ ions in and moving H^+ ions out of the thylakoid in equal numbers. Ammonium ion can also neutralize a gradient of H^+, because NH_3 can pass freely through the membrane and convert H^+ to NH_4^+.

By taking advantage of these effects and correlating them with certain optical phenomena, A. R. Crofts and D. E. Fleischman have been able to draw the following useful conclusions:

1. The intensity of delayed fluorescence from BChl in photosynthetic bacteria is a sensitive indicator of the membrane potential. Substances that dissipate the potential extinguish the delayed fluorescence. The reason is not clear; it may be that the membrane potential feeds a necessary amount of energy into the reaction that causes delayed fluorescence.

Delayed fluorescence from green plant system II is also in-

[1] Valinomycin is a doughnut-shaped molecule with a hydrophobic outside and a hydrophilic inside. The inside is just large enough to accommodate K^+ with its hydration shell. Experiments with artificial lipid membranes show that this antibiotic can carry K^+ ions through a bimolecular lipid layer.

[2] The potential is given by a Nernst equation,

$$E \text{ (volts)} = 0.06 \log_{10} \left(\frac{[K^+]_i}{[K^+]_e}\right)$$

where the subscripts i and e denote internal and external.

hibited by substances that should dissipate the membrane potential.

2. The wavelengths of the optical absorption peaks of carotenoid pigments in photosynthetic bacteria also indicate the membrane potential, and a light-induced shift of the carotenoid bands toward the red signifies the development of potential across the thylakoid membrane.[1] During illumination this potential rises to a peak of about 0.4 volt and then declines to about 0.2 volt. Similar light-induced band shifts occur in green plant membranes, but a simple quantitative relationship between the membrane potential and the amplitude of the shift has been drawn only for the bacteria.

These relationships will surely be useful as our study of the possible importance of the membrane potential progresses.

1-16. Two Hypothetical Regulating Devices

One can imagine ways in which a photosynthetic system can adjust itself for optimum satisfaction of its needs. We shall conclude this chapter by considering two examples of regulatory systems, involving membranes and/or phosphorylation, that might have some basis in reality.

1. *Reductive dephosphorylation.* The energy of ATP could be used to raise an electron to a higher energy at any point in the electron transport pattern, by moving the electron to a stronger reductant. One can easily imagine how a network such as the *z* scheme would gain in flexibility if electrons could be promoted, for example, from the top of system II to the level of NADPH. The availability of ATP and NADPH would tend to become balanced to match the needs of the organism if these two substances are linked by reactions obeying the mass-action principle; with an excess of ATP there would be more reductive dephosphorylation to form NADPH.

2. *Partition of light energy between systems I and II.* One of

[1] Not all photosynthetic bacteria show this light-induced band shift; perhaps in some varieties the carotenoid molecules are not in a proper alignment to respond to the electric field generated across the membrane.

the main conceptual difficulties with the *z* scheme has been that in its terms, sustained O_2 evolution and CO_2 assimilation require a balanced input of quanta to the two systems. The quantum efficiency for O_2 evolution is remarkably constant throughout the visible spectrum, even though there are regions where the light absorption by system II pigments (such as phycobilins and Chl *b*) is much greater than that by the Chl *a* associated with system I. This has led to the idea that excess energy absorbed by the system II pigments can "spill over" to drive system I. Efforts to measure such spillover have led to conflicting results, and it appears now that the spillover is variable and depends on the environment and its history. A few minutes of far-red illumination depresses the efficiency with which shorter-wave light can drive a system I function such as cytochrome oxidation. This treatment also increases the yield of fluorescence from the Chl *a* associated with system II. Both these responses suggest a greater apartness of the two light-harvesting systems, so that energy absorbed in the system II antenna cannot be drawn off for use in system I. A few minutes of illumination by shorter-wave (system II) light reverses this trend, apparently encouraging spillover from system II to system I as manifested by a greater efficiency for cytochrome oxidation and a lower yield of fluorescence.

The "low-spillover state" generated by far-red illumination has been called state I, and the other state II, to correspond to the kind of illumination that causes the state. The transitions between these states represent a regulatory system by which energy is distributed as evenly as possible to the two systems. The mechanism could involve swelling or shrinking of thylakoids, engendered by ion movements and leading to an actual movement of the system I and II pigments nearer to or farther from each other. If so, this would be a curious example of intrachloroplast phototaxis (see Chap. 2).

BIBLIOGRAPHY

Books (Also consult any of a variety of texts in biochemistry.)

1. Hill, R., and C. P. Whittingham: "Photosynthesis," Methuen's

Monographs, John Wiley and Sons, Inc., New York, 1955. An excellent picture of the state of research in photosynthesis in 1955.

2. Rosenberg, J. L.: "Photosynthesis," Holt, Rinehart and Winston, Inc., New York, 1965. An elementary but detailed account of photosynthesis. (Paperback.)
3. Clayton, R. K.: "Molecular Physics in Photosynthesis," Blaisdell Publishing Company, a division of Ginn and Company, Waltham, Mass., 1965. A physical approach to the problems of photosynthesis.
4. Rabinowitch, E., and Govindjee, "Photosynthesis," John Wiley and Sons, Inc., New York, 1969. A more recent treatment, with good attention to the physical fundamentals, not encumbered by too much detail.
5. "Energy Conversion by the Photosynthetic Apparatus," *Brookhaven Symp. Biol.,* no. 19, 1967. Proceedings of a symposium describing the state of research in 1966.

Reviews

6. Van Niel, C. B.: The Bacterial Photosyntheses and Their Importance for the General Problem of Photosynthesis, *Advan. Enzymol,* vol. 1, pp. 263–328, 1941.
7. Gaffron, H.: Energy Storage: Photosynthesis, in F. C. Steward (ed.), "Plant Physiology," Vol. 1B, Academic Press, Inc., New York, 1960.
8. Stanier, R. Y.: "Photosynthetic mechanisms in Bacteria and Plants: Development of a Unitary Concept," *Bacteriol. Rev.,* vol. 25, pp. 1–17, 1961.
9. Bassham, J. A.: The Path of Carbon in Photosynthesis, *Sci. Am.,* June, 1962.
10. Hind, G., and J. M. Olson: Electron Transport Pathways in Photosynthesis, *Ann. Rev. Plant Physiol.* vol. 19, pp. 249–282, 1968.
11. Greville, G. D.: A Scrutiny of Mitchell's Chemiosmotic Hypothesis of Respiratory Chain and Photosynthetic Phos-

phorylation, *Current Topics Bioenerget.,* vol. 3, pp. 1–78, 1969.

12. Levine, P.: The Mechanism of Photosynthesis, *Sci. Am.,* December, 1969.

2. Phototaxis and Phototropism: Movement and Growth in Relation to Light

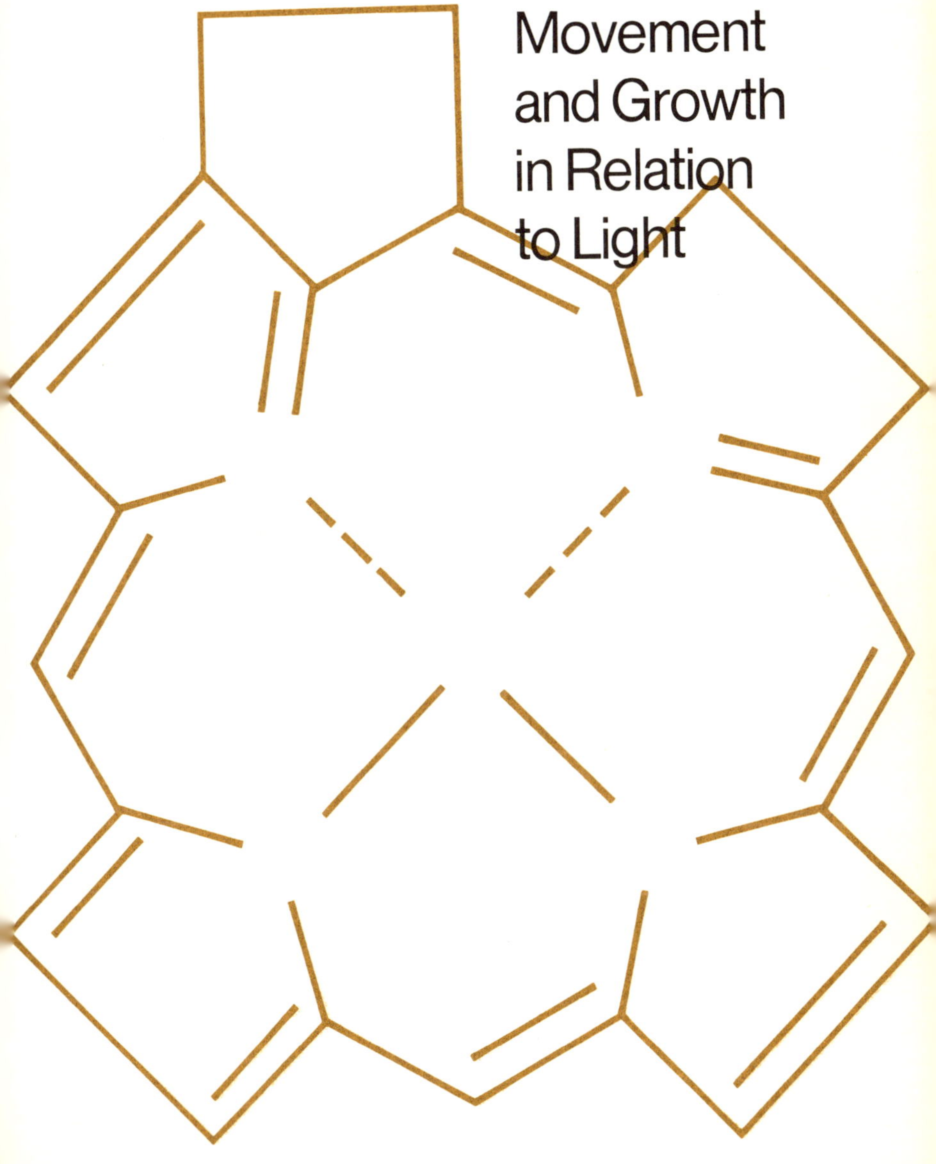

Phototaxis and phototropism are defined as free movement and directed growth, respectively, in response to light. In some cases the stimulus is a temporal change in light intensity; in others it has to do with the spatial distribution of light around or in the organism. There are countless examples of these phenomena, with a variety of apparent advantages to the light-responsive organism. Most of the examples are found in the plant kingdom, unless we include what is usually thought of as vision.

Many photosynthetic bacteria and algae exhibit phototaxis, moving so as to avoid darkness but also avoiding light intensities so high as to be damaging. The chloroplasts of algae and higher plants sometimes move within the cells, aligning themselves so that the lamellae are broadside to a weak beam of light and edgewise to a strong beam. Furthermore there might be movements of lamellae within the chloroplast to optimize the distribution of light between systems I and II (see Sec. 1-16). Seedlings of higher plants are phototropic, the tips curving toward a light source as they grow.

In all the above cases the benefit involves ongoing or future photosynthesis. In certain molds the spores are carried in balls at the ends of growing stalks; these stalks (called *sporangiophores*) grow toward the light, and the widespread dispersal of the spores is thereby enhanced.

In this chapter we shall examine a few cases of phototaxis and phototropism in enough detail to illustrate the probable mechanisms.

PHOTOTAXIS

2-1. Varieties of Phototactic Response

The phototactic responses of several microbes are illustrated in Fig. 2-1. The most beautifully simple technique is that of the photosynthetic bacterium *Chromatium*. It merely recoils upon encountering a darker region, waits about a second, and then resumes swimming. During the pause, the impact of brownian motion gives the cell a random orientation which may or may not take it back

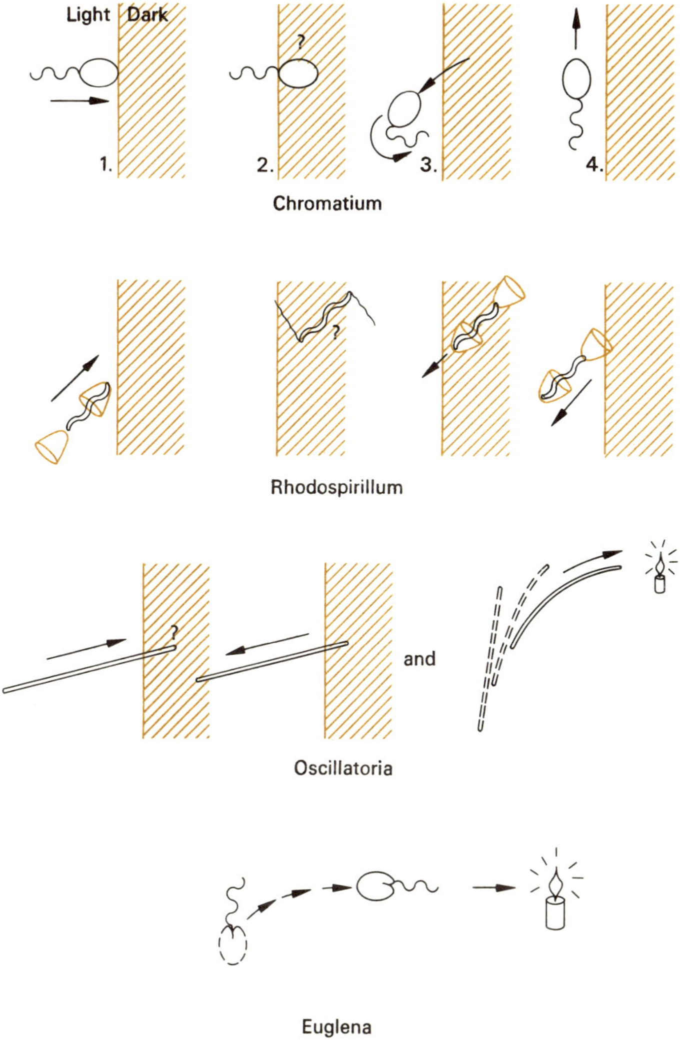

Fig. 2-1. Phototactic responses of the purple bacteria Chromatium *and* Rhodospirillum, *the blue-green alga* Oscillatoria, *and the flagellated green alga* Euglena. *Further description in the text.*

to brighter regions. If not, another response will occur when swimming is resumed.

A more coordinated and effective reaction is shown by the photosynthetic bacterium *Rhodospirillum,* which swims along its axis in either direction. If the organism experiences a sudden decrease of light intensity, as when swimming into darkness, it reverses its direction. This is achieved by reversing the orientation and sense of rotation of its flagella. The flagella, like muscle, are fibrous protein structures that can contract and thus cause motion. The detailed way of moving varies from one organism to another and is poorly understood in any case. In *Rhodospirillum* the mechanism could be that rotation of the flagella causes counterrotation of the helical cell body, which then screws its way through the water.

The reversal of movement seen in *Rhodospirillum* is exhibited also by filamentous (multicellular) blue-green algae such as *Oscillatoria,* which creep along solid surfaces and reverse the direction of creeping when stimulated by a decrease of light intensity. The filaments are covered by a polysaccharide slime, and the basis of movement might be a directed exudation of this slime by the cells.

With both *Rhodospirillum* and *Oscillatoria* it is possible to fool the cell by bringing up the boundary of a darker region from behind. Then the reversal of movement will plunge the cell deeper into darkness. This illustrates that the creatures respond simply to a temporal stimulus: to a decrease in the intensity of light reaching some receptive part of the cell, without regard to the direction from which the light has come. This kind of response is called *phobic.*

Filaments of *Oscillatoria* also show a *topic* response: one in which the direction of movement becomes orientated toward a source of light. Here the governing factor is the direction of the source in relation to the organism. This is not mediated by any local gradient of light intensity in the neighborhood of the organism. It is true that if an organism moves toward a source of light, it usually experiences an increase of light intensity. However, if a diverging lens is placed between the organism and the source, movement toward the source actually exposes the cell to progressively weaker light. Under these conditions the movement is still directed toward the source.

Flagellated green algae such as *Euglena* are structured so that their swimming can become directed toward a light source (apparent topic phototaxis) through a succession of phobic responses. This behavior involves two structures: a light-sensitive zone, or *photoreceptor,* and a bright orange spot (the *stigma*) located nearby. The photoreceptor is associated with a thickening at the base of the flagellum, and has no obvious color. Sudden shading of this photoreceptor region causes a phobic response. The cell rotates as it swims, and unless it is swimming directly toward a source of light, the photoreceptor is shaded by the stigma once every revolution. This intermittent shading can cause a succession of movements, and this might be the means by which the direction of swimming becomes aligned toward the source.

2-2. Taxis in Relation to Metabolism

From the remarkable work of T. W. Engelmann in the 1880s (see Sec. 2-3), through the extensive investigations of H. Molisch and J. Buder in the period of 1900 to 1920, and in more recent studies by A. Manten and the author it has been clear that the phototactic response in *Rhodospirillum* is closely related to photosynthesis. The action spectrum for phototaxis [1] is like that for photosynthesis. There is no special light-sensitive part of the cell for phototaxis; the response is evidently mediated by the photosynthetic light-harvesting pigments.

Rhodospirillum rubrum can grow aerobically in the dark; it then derives energy from respiration rather than photosynthesis. And in the dark, *R. rubrum* shows *positive aerotaxis,* a congrega-

[1] One can measure the action spectrum for phototaxis by setting up a split field of illumination, with a standard or reference wavelength on one side and another wavelength on the other. The intensities are then adjusted for equivalence, so that the bacteria show no tactic responses when crossing the boundary between the two wavelengths. A plot of the ratio of intensities, I (reference λ)/I (test λ), versus wavelength then constitutes an action spectrum.

tion of the bacteria in regions of greater oxygen concentration. Movement from greater to lesser oxygen concentrations elicits a reversal of swimming, as does movement from stronger to weaker light. Thus a decrease in energy-yielding metabolism is associated with a tactic response. This impression is strengthened by the competitive nature of stimulation by light and by oxygen: in strong light, which provides abundant energy from photosynthesis, the response to oxygen is absent, and conversely the tactic response to light is suppressed when oxygen is present as a source of metabolic energy. These effects are illustrated in Fig. 2-2.

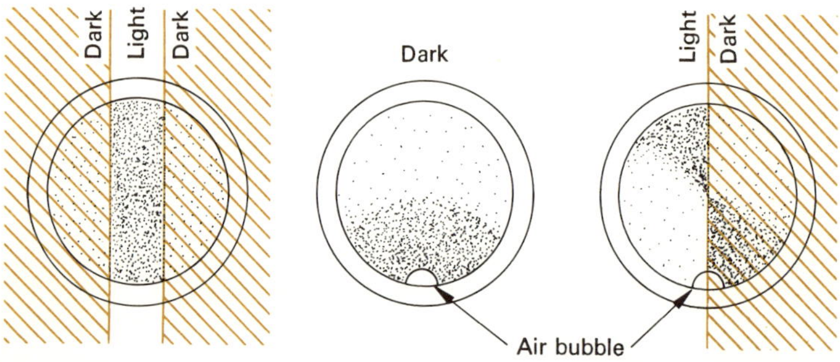

Fig. 2-2. A suspension of cells of Rhodospirillum rubrum *is held in an airtight chamber about 1 cm across and ½ mm deep. Phototaxis is evoked by projecting a strip of light onto the chamber or by darkening one half of it. Aerotaxis is produced by allowing a bubble of air to leak in. The patterns of accumulation sketched here are easily seen with the unaided eye, and illustrate that light suppresses the aerotactic response and oxygen suppresses phototaxis.*

The effects of poisons of respiration and of photosynthesis, and of other environmental factors such as pH, temperature, presence of substrates, etc., all support the idea that in *R. rubrum* a reversal of swimming direction results from a sudden decrease in the energy available through metabolism. After noting this kind of correlation in the tactic responses of a variety of organisms, J. Links suggested a general hypothesis for taxis: A response follows from a sudden decrease in the supply of energy (possibly ATP) to the locomotor or flagellar apparatus.

In blue-green algae, as in photosynthetic bacteria, the tactic responses are evidently mediated through the general metabolism. Action spectra for the reversal of creeping in *Oscillatoria* show the involvement of the photosynthetic pigments, and especially the long-wave Chl *a* that drives system I and cyclic phosphorylation (Sec. 1-9). The locomotor system (slime exudation?) is less obvious than in the photosynthetic bacteria.

Both blue-green algae and photosynthetic bacteria are more primitive than green algae in two major ways. The genetic material is not organized in a nucleus, and the photosynthetic apparatus is not organized in a chloroplast. The fact that phototaxis operates through the general photosynthetic metabolism in blue-green algae and photosynthetic bacteria again suggests a primitive status. In green algae there are specialized structures and pigments, separate from the chloroplasts and their pigments, for phototaxis.

The green alga *Euglena* (see earlier) has a photoreceptor for taxis near the base of the flagellum, and nearby an orange stigma containing carotenoid pigments. Action spectra for phototaxis show in general that blue light is effective and red light is not. This rules out chlorophylls and draws attention to carotenoids and flavins (Fig. 2-3) as possible active pigments in the photoreceptor. Interpretation of the action spectra is confusing, because the tactic response involves the ability of the stigma (and of the chloroplast, for that matter) to shade the photoreceptor. The effectiveness of light at any wavelength therefore involves the absorption spectra of the shading pigments as well as the photoreceptive pigments. This problem can be sorted out with the help of mutants that lack the stigma, the chloroplasts, or both.

In normal *Euglena* the action spectrum for phototaxis shows a principal maximum at 495 nm, a secondary one at 425 nm, and no action beyond about 530 nm. But in a mutant that has neither chloroplasts nor stigma the action spectrum shows a peak at 410 nm and diminishes smoothly toward greater wavelengths, reaching

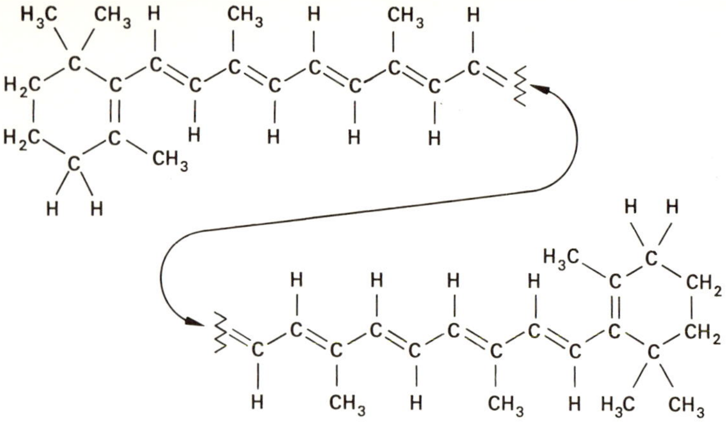

beta-Carotene

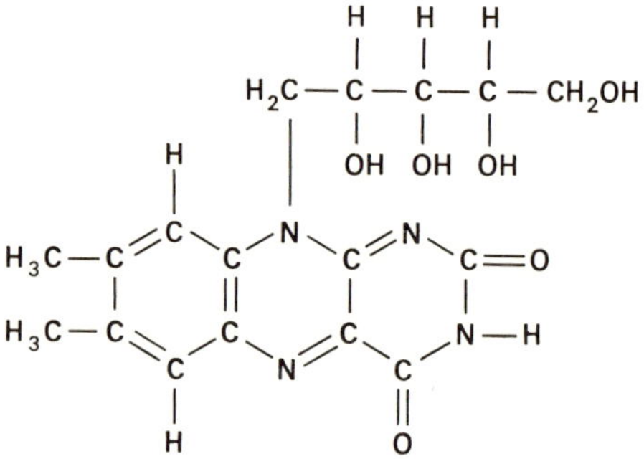

Riboflavin

Fig. 2-3. Structural formulas of β-carotene and riboflavin. The carotenoid pigments function in photosynthesis and in protection against damage by light as discussed in Chaps. 1 and 5. Flavins, which can undergo oxidation and reduction, function as carriers in electron transport.

zero at about 530 nm. The action spectrum for the mutant probably reflects more cleanly the absorption spectrum of the receptor for phototaxis but does not give us a clear identification of the receptor pigment(s).

If the phototaxis of *Euglena* and other green flagellates is to be brought under Links' hypothesis, we must assume that the receptor mediates a change in the supply of energy to the base of the flagellum without the obligatory involvement of photosynthetic phosphorylation.

2-3. T. W. Engelmann: An Appreciation

Much of our present knowledge of the roles of pigments in photosynthesis and in phototaxis rests on the imaginative and artful work of T. W. Engelmann, reported mainly in the 1880s. Three examples will show the character of this work:

1. While observing a single cell of *Euglena* through a microscope, Engelmann was able to induce tactic responses by shading small regions within the cell. In this way he showed that the light-sensitive zone is at the base of the flagellum and is not the highly suggestive stigma.

2. By projecting a microspectrum into the field of a microscope, onto a suspension of photosynthetic bacteria, Engelmann showed that as a result of phototaxis the bacteria congregate mainly in the near infrared between 800 and 900 nm. He inferred that this was a region of absorption by a pigment (bacteriochlorophyll) important for photosynthetic growth of the bacteria.

3. Projecting a microspectrum along a single filamentous alga, Engelmann expected that in those regions where the wavelength was favorable for photosynthesis, the oxygen concentration would be greater. To test this he introduced a population of aerotactic bacteria around the filament and let the preparation rest in the dark until it had become free of oxygen as a result of respiration by the bacteria. Then when the illumination (in the form of a spectrum) was turned on, the bacteria began to congregate, by aerotaxis, around those parts of the filament that were evolving oxygen the most rapidly. The bacteria thus assembled themselves so as to display the action spectrum for photosynthetic oxygen evolution by the algal filament. Engelmann thus found that light in the region of 570 to 630 nm, absorbed mainly by phycobilins

rather than cholorophylls, was particularly effective for photosynthesis. This provided the first evidence that accessory pigments, other than chlorophylls, could function in photosynthesis.

PHOTOTROPISM

2-4. General Characteristics of Phototropic Systems

The growth of many plants shows a bending or turning toward the light. This phototropism has been studied most intensively in oat seedlings and in the mold *Phycomyces*. In oats the object of study is the coleoptile, or growing tip of the seedling. In *Phycomyces* it is the sporangiophore, the stalk that supports the spherical sporangium which holds spores prior to their release.

The sporangiophore of *Phycomyces* grows to a height of several centimeters, at a rate of about 3 mm/h. It has a diameter of about 0.1 mm near the top and supports a sporangium about 0.5 mm in diameter (Fig. 2-4). The active tissue forms a shell around a central watery vacuole, probably with two membranes and a cell wall. New growth and phototropic sensitivity are both confined to the uppermost 2 mm (approximately) below the sporangium. In the oat coleoptile the sensitivity to light is greatest at the tip, above the major part of the growing zone.

When illuminated from the side, these cylindrical structures (the oat coleoptile or the sporangiophore) bend toward the source of light as they grow, as shown in Fig. 2-5. The bending happens because the back side of the stalk (away from the light) grows faster than the front side. At first sight it might seem that the front side receives more light and the faster growing back side is relatively shaded. This may be the case with some seedlings, but in *Phycomyces* the sporangiophore is nearly transparent and corresponds optically to a cylindrical lens (Fig. 2-6*a*). Parallel rays entering at the front converge and produce a zone of greater intensity at the back. It appears therefore that the distribution of light is important in setting the growth rate, with faster growth in regions

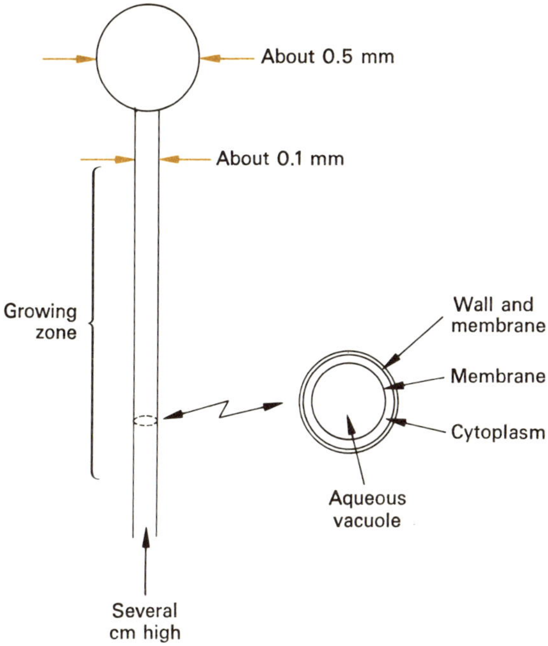

Fig. 2-4. The sporangiophore (stalk) and sporangium (ball containing spores) of the mold Phycomyces. *The growing zone, about 2 mm long near the top, is also the light-sensitive region for phototropic stimulation.*

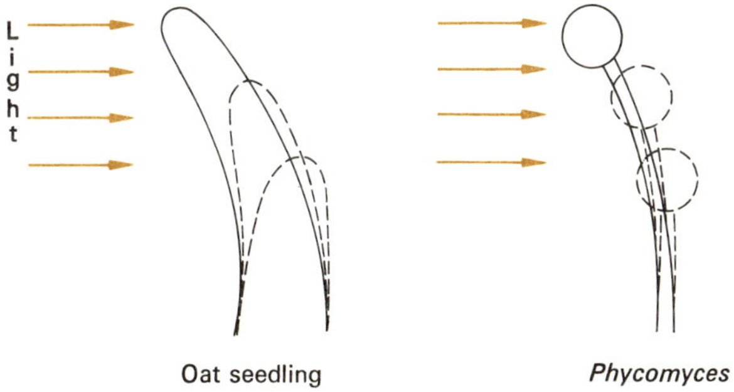

Fig. 2-5. Phototropic curvature in the growth of the oat seedling and the Phycomyces *sporangiophore.*

of greater intensity. This view is supported by experiments in which the distribution of light is changed:

1. The sporangiophore can be immersed in a medium of greater index of refraction, so that the rays entering the cylindrical structure diverge instead of converging (Fig. 2-6*b*). The plant then grows with a slight curvature away from the light, as befits a higher intensity on the front surface.

2. The interior of the *Phycomyces* sporangiophore contains gallic acid, which absorbs ultraviolet (around 280 nm) strongly. The phototropic response to ultraviolet is negative (curvature

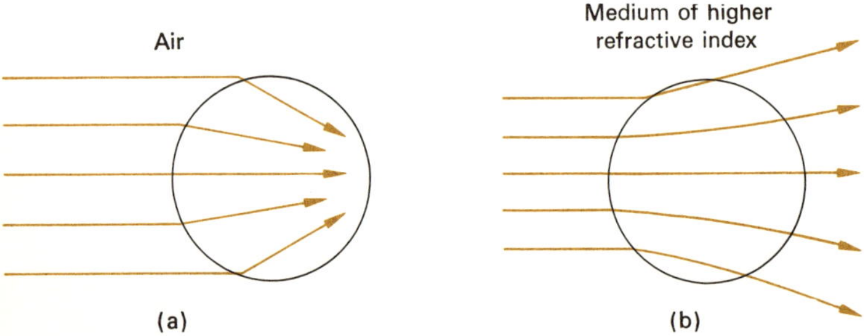

Fig. 2-6. The sporangiophore of Phycomyces *behaves like a cylindrical lens. (a) In air, parallel rays entering the structure converge to a focus near the far side. (b) If the structure is immersed in a medium of greater refractive index, it acts like a diverging lens.*

away from the light), even for sporangiophores in air, because the rear is shaded by the gallic acid.

3. *Phycomyces* shows a light-growth response: uniform illumination (from all sides) allows straight vertical growth, but if the intensity of this uniform light is suddenly raised, there is a transitory spurt of growth. A drop in the intensity causes a temporary slackening of growth. Faster growth is thus associated with greater intensity of illumination.

The light-growth response has been studied exhaustively, especially in *Phycomyces* by E. S. Castle and M. Delbrück and their collaborators. A remarkable feature of this response is the adaptation. After a step increase (or decrease) of intensity there is a transitory increase (or decrease) in the growth rate, but after a few minutes at the new intensity the rate returns to what it was before. Thus the intensity can be changed over many orders of magnitude with no long-term change in the growth rate. This poses a riddle: If the tropic response is simply a special manifestation of the light-growth response, why does adaptation not equalize the effects of the more concentrated light on one side and the more diffuse light on the other? How can curved (differential) growth be sustained beyond the time needed for complete adaptation? L. Jaffe has suggested that an adapting substance is generated in the tissue by light, and can diffuse away from the illuminated regions. Because of this diffusion, the state of adaptation in the regions of higher light intensity might lag perpetually behind that in the dimmer regions, allowing a difference in growth rates to persist.

The phenomenology of phototropism is far more complex in higher plants than in *Phycomyces*. Both the phototropic and the light-growth responses of seedlings show complicated dependence on the history and geometry of illumination. In contrast to the case of *Phycomyces,* it does not seem appropriate to interpret the phototropism of a seedling as a special case of the light-growth response. For example, cutting off the tip of the growing oat coleoptile destroys the phototropic response while still allowing a light-growth response.

If just one side of the oat coleoptile is illuminated from above while the other side remains shaded, the plant shows a phototropic curvature *toward* the illuminated side (not away from it, as *Phycomyces* would behave). Growth is faster on the shaded side. If diffusion of matter from one side to the other is prevented by insertion of a thin glass plate, there is no phototropic response. Thus it may be said that light generates a substance which must diffuse across the growing coleoptile in order to have an effect on the opposite side. The effect would have to be a stimulation of growth to account for the direction of curvature.

The growth of plants is subject to control by hormones such as indoleacetic acid (IAA, or auxin). Thus it was natural for N.

Cholodny and F. Went to propose, in the 1920s, that phototropism is mediated by auxin. Light, striking the tip of the oat seedling, might cause destruction of auxin and therefore a slowing of growth on the illuminated side or else a movement of auxin to the opposite side, where growth is promoted (the latter view became favored by the experiment with the glass plate). Now, after decades of experimentation, it must be said that the presence of auxin is needed for phototropism (as it is for growth), but direct involvement in the phototropic mechanism is neither proved nor disproved. This state of affairs is representative of our knowledge of molecular mechanisms in phototropism.

2-5. The Problem of the Photoreceptor

A problem that is encountered repeatedly in photobiology is that of a photoreceptor pigment, apparently absorbing in the blue or violet, but unidentified for several reasons: (1) Its concentration in the tissue is so small that its optical density is about 10^{-3} or less, making direct absorption measurement impractical. (2) The information that can be inferred about the absorption spectrum, for example from the action spectrum, is not distinctive enough to allow unequivocal identification. (3) The action spectrum is so distorted by screening pigments that it does not reflect the absorption spectrum of the photoreceptor pigment. This lack of knowledge holds for phototaxis in green algae and for phototropism in plants and fungi. We can only guess that the receptor might be a carotenoid, or a flavin, or a pterin, etc.

Some results of action spectrum measurements in various phototactic or phototropic systems are tabulated below. Activity by light beyond about 530 nm (as for the phototaxis of the dinoflagellate *Prorocentrum*) is uncommon.

Upper and lower limits have been estimated for the concentration of the photoreceptor pigment in *Phycomyces*. The optical density of a sporangiophore, along a path perpendicular to its axis, is less than about 0.01 in the visible region for any possible receptor pigment. The path length through the cytoplasm is about

Table 2-1. Maxima in the Action Spectra for Phototaxis or Phototropism of Certain Organisms.

	Maxima in the action spectrum		
	Principal	*Secondary*	*Ultraviolet (usually not measured)*
Phototaxis in flagellated green algae:			
Platymonas	493	435	340, 270
Ulva (gametes)	485	435	
Gonyaulax	475	None	
Prorocentrum	570	None	
Euglena, normal	495	425	
Euglena, mutant without stigma or chloroplasts	410	None	
Phototropism:			
Avena (oat) seedling	475, 445	425	370
Phycomyces	485, 450	420	380; negative at 280*

* Due to screening by gallic acid.

3×10^{-3} cm, and the extinction coefficient for a pigment such as a carotenoid is about $3 \times 10^{4} M^{-1}\,cm^{-1}$ at the absorption maxima. Therefore the relation

$$\mathrm{OD} = \varepsilon Cx \qquad (2\text{-}1)$$

(see Sec. 3-3, Volume 1), together with the stipulation OD   0.01, can be written

$$0.01 \gtrsim (3 \times 10^{4}) C (3 \times 10^{-3})$$

or $C \lesssim 10^{-4} M$.

On the other hand the concentration cannot be much less than $10^{-5}M$. If it were, the rate of quantum absorption would be too low to account for the minimum or threshold intensity for phototropic curvature. It is necessary that within the time that elapses between stimulation and response (a few minutes), enough quanta be absorbed to assure a distinction between the back and the front (relative to the source of light) of the sporangiophore. For a pigment of extinction coefficient about $3 \times 10^4 M^{-1}\ \text{cm}^{-1}$, distributed in a light-sensitive zone about 2 mm long in the cytoplasm of the sporangiophore, the concentration must be about $10^{-5} M$ or more to meet this requirement at the observed threshold intensity. The corresponding OD, measured perpendicular to the sporangiophore axis, should be about 10^{-3}.

Attempts have been made to detect transient light-induced changes in the OD of the sporangiophore, in the hope that the photoreceptor pigment undergoes a photochemical cycle that is manifested by a change in absorption spectrum. Such attempts, with equipment able to detect changes of 10^{-4} OD lasting about 0.1 sec or more, have not been successful.

MECHANISM

2-6. Speculations about Mechanism

We know almost nothing about the molecular mechanisms of phototaxis and phototropism. A few loosely related facts can be listed, to encourage speculation.

1. Links' hypothesis: A phobic tactic response can usually be correlated, at least in principle, with a decrease in the supply of energy to the motor system.

2. Coupling: The link between electron transport and ATP formation involves the controlled movement of ions across membranes.

3. Phototactic behavior in certain green algae is governed by the concentrations of Ca^{++}, Mg^{++}, and K^{+} in the medium. Specifically, P. Halldal found that *Platymonas* in a low-K^{+} medium swims toward a source of light if the ratio $[Mg^{++}]/[Ca^{++}]$ is greater than 6, swims away (negative phototaxis) if this ratio is less than 6, and swims in a random direction if the ratio $[Mg^{++}]/[Ca^{++}]$ equals 6.

4. The concentrations of Mg^{++}, Ca^{++}, and K^{+} are influential in determining ATPase activity and movement in contractile

systems including muscle and flagella. In muscle the contraction is mediated by a change in [Ca^{++}], initiated by a nerve impulse and regulated by a membrane.

5. Isolated cilia and flagella undulate, as shown by H. Hoffmann-Berling, in the presence of ATP; the frequency of undulation rises with increasing ATP concentration.

6. Membranes can undergo extensive changes in their electrical polarization and their permeability toward chemicals, in response to small chemical or physical disturbances. This is seen in nerve excitation, where a current through a small part of the nerve membrane initiates a self-sustaining wave of ionic current that travels along the entire nerve.

It is not difficult to assemble these facts in a sketchy hypothesis for the mechanisms of tactic and tropic responses: Light, through a photochemical process, causes a chemical perturbation at a certain membrane. A change in the property of this membrane initiates a tactic or tropic response, in one way or another. For taxis involving movement by means of flagella, a change in concentrations of cations and/or ATP causes a change in the pattern of flagellar contraction. For *Oscillatoria* the exudation of slime (for creeping) might be controlled by a membrane, and in phototropic systems a membrane could regulate the flux and the local concentrations of growth hormones or metabolites.

These are nothing more than idle speculations, but they may indicate the desired direction of future biochemical studies.

BIBLIOGRAPHY

1. Clayton, R. K.: Phototaxis in Micro-organisms, in A. C. Giese (ed.), "Photophysiology," Vol. II, pp. 50–77, Academic Press, Inc., New York, 1964. A detailed and annotated review of phototaxis in photosynthetic bacteria and algae.
2. Ball, N. G.: Tropic, Nastic and Tactic Responses, in F. C. Steward (ed.), "Plant Physiology," Vol. 5A, pp. 119–228, Academic Press, Inc., New York, 1969. Reviews phototropism quite thoroughly.

3.
Vision in Higher Animals

Of all the senses, vision is unique in its ability to provide an organism quickly with a detailed map of the near and far surroundings. The way that this is done seems related to the needs of an animal. For example, predators such as owls and cats have their eyes facing forward, and traditional prey such as pigeons and rabbits have their eyes on the sides of their heads. The predators can see where they are going, and the prey can see what is coming. More subtle specializations are found in the way that visual information is handled: some animals, such as cats, respond principally to form, whereas others, such as pigeons, are more responsive to sudden movements.

Except in its most primitive forms [1] an eye is like a camera, in that the visual field is mapped as an image on a surface (the *retina*) made up of light-sensitive cells. This mapping is achieved in at least three distinct ways, sketched in Fig. 3-1. The simplest image-forming eye, found in some molluscs, is analogous to a pinhole camera. Because of its small aperture it can admit relatively little light. A second basic plan is found in the many-faceted compound eyes of insects and spiders. Here the light from a certain direction can enter only those channels that point in that direction. Finally, there is the system most like a conventional camera, using a lens to form an image. This system, found in vertebrates, will receive most of our attention.

Seeing begins with the absorption of light by a pigment in the receptor cells of the retina. The pigment undergoes a cycle of changes, and an electrical effect ensues: an ion current flows, and a voltage develops across the receptor cell membrane. Here the effect of one light quantum is amplified into the movement of about 10^5 sodium ions. The electrical effect in the receptor cell is communicated through a network of several kinds of nerve cells, leading to an impulse (a traveling wave of transmembrane ion current) in a fiber of the optic nerve, which goes to a part of the brain. These events involve an intricate interplay of information from

[1] Primitive "visual" systems include the receptors for phototaxis in algae (see Chap. 2) and also the light-sensitive cells or clusters of cells found on the surfaces of some simple animals such as worms and molluscs.

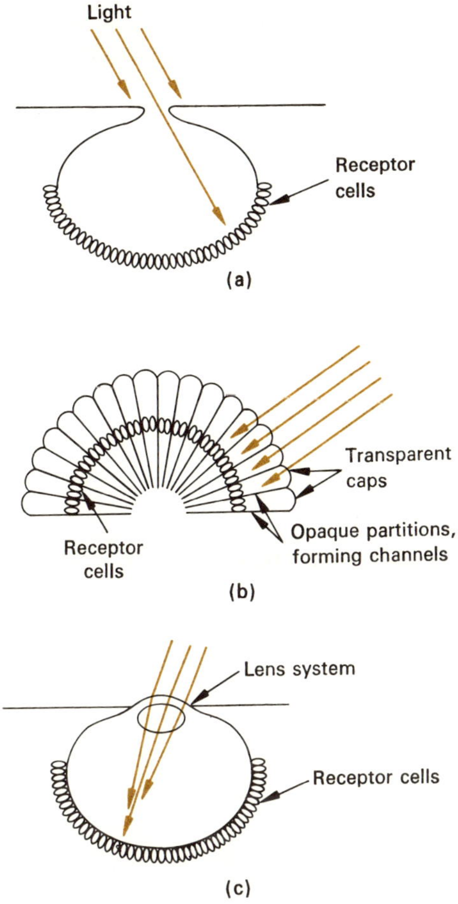

Fig. 3-1. Three ways that the eye of an animal can form an image on a layer of light-sensitive receptor cells. (a) The "pinhole camera" plan of some simple marine animals; (b) the compound eye found typically in insects; (c) the cameralike plan of the vertebrate eye.

the various receptor cells and through the various nerve paths. As a result the animal discriminates forms, movements, and (in some cases) colors.

In studying vision we will therefore attend to the following aspects:

1. Anatomy: gross, cellular and subcellular
2. Factors that determine visual acuity and sensitivity; dark adaptation
3. The chemistry and photochemical cycle of the visual pigmer ts
4. The basis of color vision
5. The amplifying transduction between optical input and electrochemical output in a receptor cell
6. Mechanisms in the processing of information by the nerve network

ANATOMY AND FUNCTION IN THE EYE

3-1. Anatomy

The human eye is diagrammatically sketched in cross section in Fig. 3-2. Light entering the eye is deviated so as to form an image. In land animals most of this refraction happens at the cornea, where the discontinuity in the index of refraction is greatest. The lens provides a secondary focusing adjustment; its convexity can be varied by the ciliary muscle.

The iris, a circular muscle of variable aperture, contracts in strong light. Thus when the level of illumination allows a smaller window in the eye, the iris ensures that only the optically best center part of the lens is used. This is partly responsible for the improved ability to see fine details in stronger light. Another factor in visual acuity, emphasized by A. Rose, is that the arrival of light quanta at the receptor cells is random. This randomness, or "quantum noise," is less significant at higher intensity.

The iris also gives some control over the amount of light entering the eye and helps prevent a dazzled sensation in strong light.

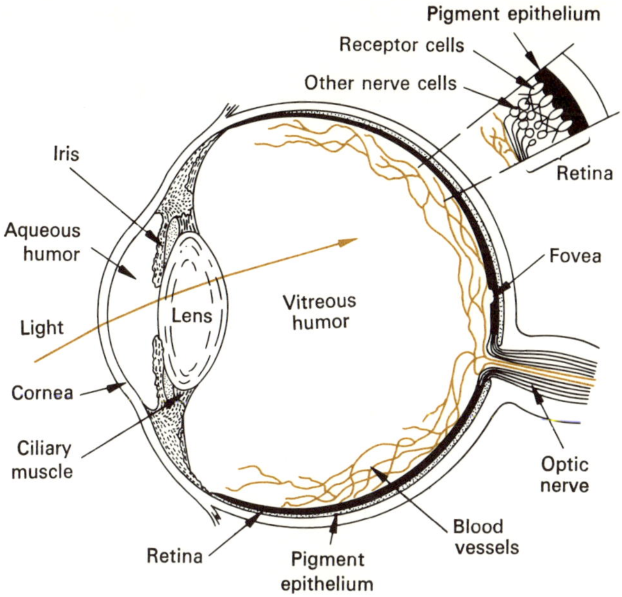

Fig. 3-2. A diagrammatic view of the human eye. The pupil (not shown) is the image of the iris as seen from outside the eye. The various structures are described in the text.

The back of the eye has a layer of light-sensitive receptor cells set in a bed of pigmented epithelium. The black pigment in the epithelium, like the blackening of the inner surfaces of a camera, serves to reduce the haze of scattered light in the eye. This black substance is not the visual pigment, which is in the receptor cells.

The receptor cells are covered by a layer of nerve cells, optic nerve fibers, and blood vessels, through which the light must pass to reach the receptors. This can be regarded as an error in the evolution of the vertebrate eye. The seriousness of the error is alleviated in a small region that is relatively free of obstructing tissues. This central part of the retina, called the *fovea,* provides the crisp vision that we have when we fixate on an object.

The optic nerve fibers pass out of the eye in a bundle where there is no room for receptor cells. This "blind spot" is toward the nose from the fovea. It can be found as shown in Fig. 3-3.

Fig. 3-3. Find the blind spot, where the optic nerve leaves the eye, as follows: Hold your hand over your left eye, and fixate on the cross with your right eye. If you hold the page upright about 8 in. from your eye, the dot should be invisible.

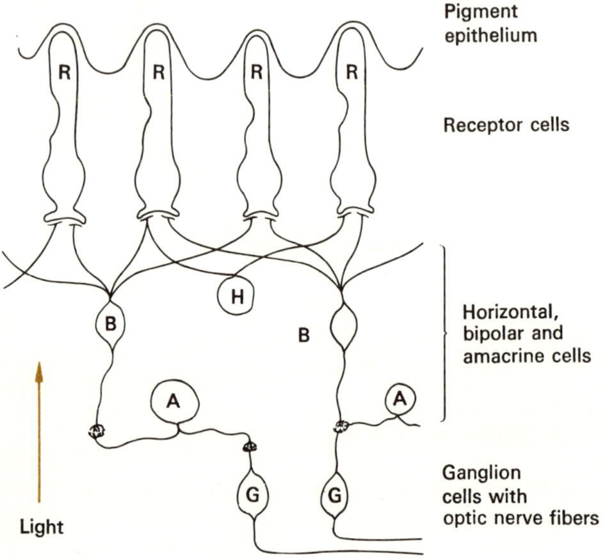

Fig. 3-4. A schematic drawing of the types of nerve cells and interconnections in the retina, between the light-sensitive receptor cells and the optic nerve.

The classic elucidation of the anatomy of the retina was by S. Ramon y Cajal, about 1900. This was extended three decades later by the extensive work of S. L. Polyak (see the Bibliography) and G. Østerberg. There are five types of nerve cells in the retina: receptors, horizontal cells, bipolar cells, amacrine cells, and ganglion cells. They form a network of communication between the surroundings and the brain, processing the visual information and delivering it through the optic nerve. The types of cells and their interconnections are shown schematically in Fig. 3-4. Electrochemical signals from the receptor cells are picked up by the bipolar cells, which in turn stimulate the ganglion cells (either directly or by way of amacrine cells). The optic nerve fibers are outgrowths of the ganglion cells; stimulation of these cells causes volleys of impulses to pass down the optic nerve to the brain. There is usually convergence, in that one ganglion cell can receive signals originating in many different receptors (see the next section). The horizontal cells establish cross connections among receptor and bipolar cells, and the amacrine cells provide contacts between bipolar and ganglion cells. The thin filaments going from one nerve cell to another are made evident only by special stains; a cross section of the retina presents the appearance of dense layers of consecutive types of nerve cells sandwiched together.

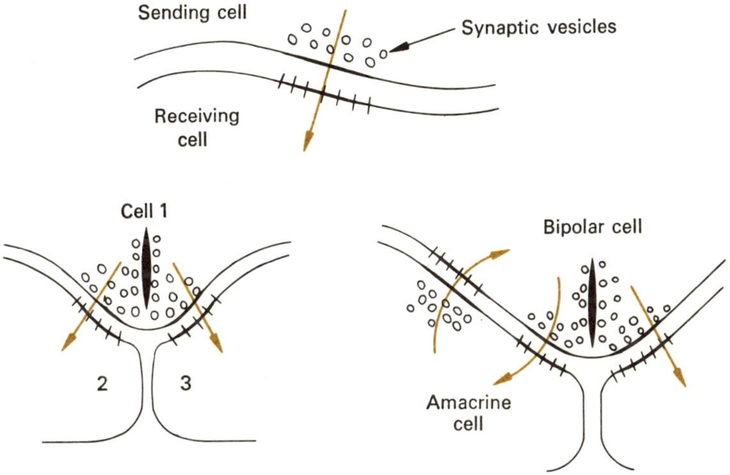

Fig. 3-5. A synapse is a region of contact where a signal is transmitted from one nerve cell to another. In the electron microscope, synapses look somewhat like these drawings (the directions in which signals are transmitted are shown by arrows). Receptor cells and bipolar cells often form the sending units of double synapses (center). The boundary between a bipolar and an amacrine cell often shows a pair of synapses of opposite polarity, suggesting a feedback loop (bottom).

Areas of contact where information is transferred from one nerve cell to another are called *synapses.* When thin slices of retina are viewed in the electron microscope, the synapses look like the sketches in Fig. 3-5. The cell membranes appear denser at a synapse, and there are ball-like and fiberlike structures within the sending and receiving cells respectively (polarity is indicated by the arrows). Sometimes the sending cell has what looks like two synapses squeezed together, with signals being transmitted from that cell to two others. This double synapse is seen only in receptor cells and bipolar cells. In amacrine cells one can often find closely

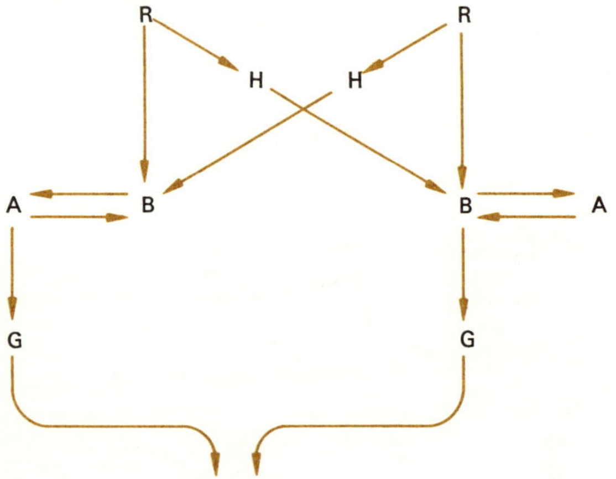

Fig. 3-6. Major pathways from receptor cells (R) through horizontal, bipolar, and amacrine cells (H, B, A) to ganglion cells (G) and on to the brain. The significance of these pathways is discussed in Sec. 3-10.

spaced synaptic regions of opposite polarity, suggesting local "information feedback loops."

The patterns of synaptic connections have been studied in the retinas of a variety of animals, and with particular success by J. E. Dowling and collaborators in the retina of the mud puppy, a salamander in which the nerve cells are unusually large. The most common pathways, already suggested by Fig. 3-4, are summarized in

Fig. 3-6. The significance of these pathways will be examined in Sec. 3-10.

The vertebrate retina contains two major classes of receptor cells called *rods* and *cones*. They have distinct properties and functions, as we shall see.

3-2. Rods and Cones

The vertebrate rod (see Fig. 3-7) is an elongated cylindrical cell; a narrow waist divides it into outer and inner segments. The outer segment measures about 1 $\times$ 10 microns in man and about 6 $\times$ 60 microns in the frog. It contains several hundred thylakoids, or

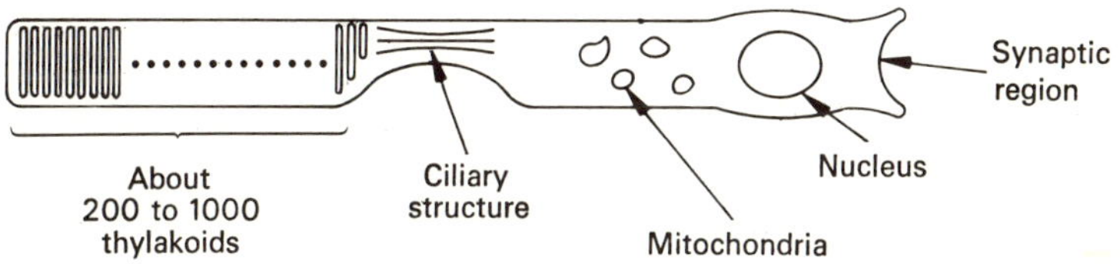

Fig. 3-7. A vertebrate rod. The length of the outer segment ranges from about 10 microns in man to about 60 microns in the frog. The sizes of the thylakoids are exaggerated; actually they are much thinner (about 200 Å) and more closely spaced.

flat membrane-bound bags, each about 200 Å thick, that support the visual pigment. The inner segment contains mitochondria (structures that perform energy metabolism), a nucleus, and synaptic regions at the far end. A ciliary structure can be seen at the waist between the segments; in fact the rod develops embryologically from two cilia.

The principal visual pigment is rhodopsin, a compound of the chromophore [1] *retinene,* or *retinal,* and the protein *opsin* (see Sec. 3-5 and Fig. 3-15).

The transition dipole moment of retinal is along the chain of conjugated double bonds, so the absorption of light is most effective

[1] A chromophore is that special part of a pigment molecule which is responsible for the absorption of light. It should not be confused with a chromatophore as defined in Chap. 1.

when this chain is parallel to the electric vector of the light wave, perpendicular to the direction of propagation of the light. With the conjugation chains of the retinal molecules confined to the planes of the thylakoid membranes, perpendicular to the direction of incoming light, the rod achieves 50 percent more light absorption than with an arrangement that is random in three dimensions. Within the plane of the membrane, the direction is random in the eyes of vertebrates.[1] The molecules of rhodopsin in the thylakoid are about 50 Å apart, and there are about 10^7 molecules in a single rod. This large number is needed in order for a rod to have a reasonable chance of absorbing an incoming quantum: at the wavelength of maximum absorption, 500 nm for rhodopsin, the optical density of a rod in the axial direction is about 0.1 to 0.3, corresponding to 10 to 50 percent absorption.

The thylakoid membrane is made almost entirely of rhodopsin, intercalated with some phospholipids. More than 60 percent of the membrane protein is opsin.

Although the rhodopsin in a rod is confined to the membrane of an internal structure, this membrane comes very close at its edges to the outer membrane of the cell. In the other kind of vertebrate visual receptor, the cone, the pigment is in the external membrane itself. Here the external membrane is folded in repeatedly, forming a structure somewhat like the stack of thylakoids in a rod. In fact one can speculate that the thylakoid structure in the rod represents a further development of the folded membrane of the cone, as suggested in Fig. 3-8.

In contrast to the vertebrate rods and cones, the squid photoreceptor has thousands of tiny fingers, or microvilli, sticking out of a cylindrical cell (again see Fig. 3-8). These microvilli serve to expand the surface of an external membrane that supports the visual pigment. The squid retina lacks intermediate nerve cells; the optic nerve fibers are direct outgrowths of the receptor cells.

[1] In some insects, such as bees, the arrangement of rhodopsin in the membrane is orderly, and the eye can distinguish the direction of the electric vector when the incoming light is polarized. Since sunlight scattered from the sky is partly polarized, the bee can locate a patch of sky with reference to the position of the sun even when the sun cannot be seen.

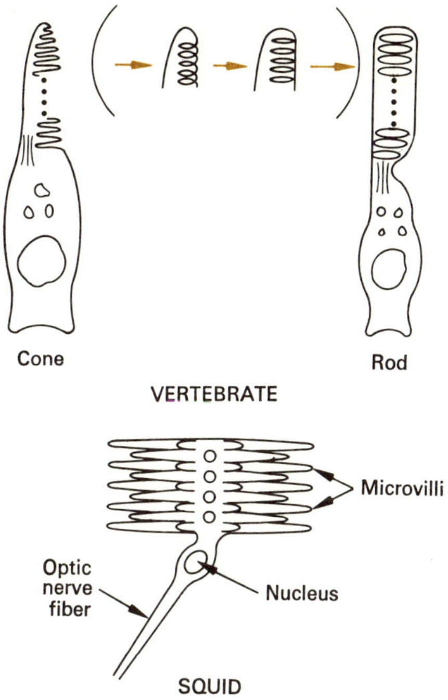

Fig. 3-8. A vertebrate cone, a vertebrate rod, and a receptor cell in the retina of the squid. Possible stages in the development of the internal thylakoids of rods are shown. The sizes of repeating structures (membrane layers and microvilli) are exaggerated; actually they are far more numerous and much thinner.

Not only are rods and cones constructed differently, but they contain slightly different visual pigments (see Sec. 3-5), they are disposed differently in the retina, and they serve different specialized functions in seeing.

The human fovea, the fixation point at the center of the field of vision, is about 0.3 mm in diameter. It contains about 10,000 cones and no rods. The cones are about 2 to 5 microns apart, and their nervous connections to the ganglion cells show little convergence: for the most part each cone is connected, through a bipolar cell, to a separate ganglion cell. This makes for high acuity, since the stimulation of two adjacent cones can lead to activity in separate fibers of the optic nerve.

Altogether there are about 6 million cones in the retina, but their density is about twentyfold less outside the fovea and diminishes steadily toward the periphery. In these peripheral regions the rods (about 125 million in all) predominate; here their density of $150{,}000/mm^2$ rivals that of the cones in the fovea.

Despite their close spacing, the rods do not provide high acuity, as the foveal cones do. Signals from as many as 10,000 rods converge, through the horizontal, bipolar, and amacrine cells, to a single ganglion cell. In some convergent pathways the signals are antagonistic and tend to cancel each other, and in others they are additive (see Sec. 3-4). Regardless of these interactions the limit of resolution with peripheral vision is about 100 rods: any contrast of light and shadow within a field of that size cannot be appreciated.

What the rods lose in acuity they gain in sensitivity, by pooling the signals of many at a common point. But the greater sensitivity of rod vision compared to cone vision cannot be understood solely on the basis of convergence of the nerve net. The image of a star is essentially a diffraction image that covers only a few receptor cells (see the next section). To see a faint star with the dark-adapted eye one should look a little away from it so as to use the periphery instead of the fovea. Here the effect of convergence is missing, and still the visual impression made by a few rods is much greater than that made by a few cones. We shall return to this problem in Sec. 3-4. For the present, it suffices to say that the rods are far superior to the cones for seeing in very weak light after prolonged dark adaptation but in strong light the rods become less sensitive than the cones.

The cones are specialized for color vision as well as for high acuity. In man this delightful sense is based on the existence of three different cone pigments, as will be discussed in Sec. 3-8. Color vision is found mainly in the fovea and in the more central parts of the peripheral retina (just ouside the fovea), where the cones are relatively numerous. The improvement in color vision, going from the periphery of the retina to the fovea, reflects the increasing abundance of cones. To see this improvement, hold an object of bright but unknown color (such as a crayon selected without looking) just behind the field of vision. While fixating straight ahead, bring the object slowly toward the center of your field of view, and notice when you can identify the color with confidence.

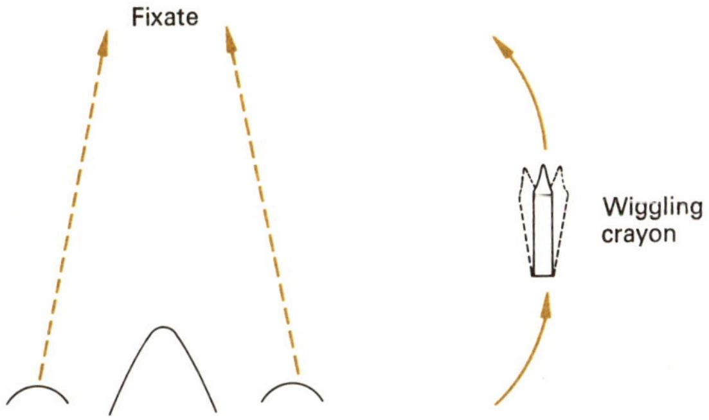

Fig. 3-9. The periphery of the retina responds better to movement than to form, and is deficient in color vision. To see this choose a colored crayon from behind your back, fixate straight ahead, and move the crayon slowly, while wiggling it, into your field of view. You will detect movement first, then form and color.

The same test, with the modification that the crayon is wiggled as it is moved into the field of view, will show that the most peripheral part of the retina responds to movement more than to form (Fig. 3-9). You will get the flickering impression that something is there before you can actually "see" the object. This is probably a function of neural organization (see Sec. 3-10) rather than rods versus cones.

THE LIMITS OF VISUAL ACUITY AND SENSITIVITY

3-3. Visual Acuity

Visual acuity can be described by the ability to distinguish two closely spaced objects: dark lines or dots on a light background, or point sources of light against a dark field. Considering two points that are barely resolved, one can specify their angular separation (the angle subtended at the eye) or the corresponding lateral distance at the retina.

Any one of several factors can set the limit of acuity: the spacing of receptor cells having independent nerve paths, the effect of diffraction, and the various defects (aberrations) in the formation of images by lens systems.

In man the closest spacing of neurologically independent receptors is that of cones in the fovea, about 2 to 5 microns center

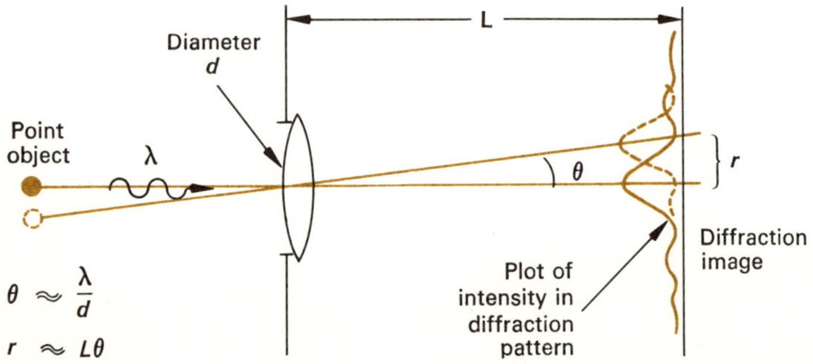

Fig. 3-10. Because of the wave nature of light, a lens translates a point object not into a geometrical point image but into a diffraction pattern. The approximate radius r *of the pattern is determined by the equations shown.*

to center. Since the receptor cell becomes excited as a unit, one cannot hope for better resolution than 2 microns at the retina. If the retina is about 2 cm, or 2×10^{-4} micron, from the cornea, this corresponds to an angle of about 10^{-4} radian between the incoming rays to two points that can be resolved. Projecting this angle into object space gives a resolving power of about 1 in. at a distance of 1,000 ft.

In one way or another the effect of diffraction dictates that one cannot resolve details smaller than the wavelength of the light that is used for seeing. With a lens, a point object is translated, not into a point image, but into a diffraction pattern, the result of constructive and destructive interference of the light waves going through the lens. The pattern is a central bright spot surrounded by concentric, ever fainter, dark and light rings. The distance between the center and the first dark ring can be taken roughly as the radius of the diffraction pattern. For two points to be well resolved, the diffraction patterns of their images should be separated by at least this radius. The smaller the aperture of the lens, the larger the diffraction pattern; an approximate relationship (see Fig. 3-10) is

$$\Theta \approx \frac{\lambda}{d} \tag{3-1}$$

where λ is the wavelength of the light, d is the diameter of the lens aperture, and Θ is the angle subtended at the lens by the radius of the pattern. In the eye the aperture d is the diameter of the pupil. If the retina is a distance L from the cornea (the effective lens) and the radius of the pattern is r, then $\Theta \approx r/L$. With $\lambda = 500$ nm, or 0.5 micron, and $d = 5$ mm, Eq. (3-1) gives $\Theta = 10^{-4}$ radian, and if $L = 2$ cm, $r = 2$ microns. Thus for a pupillary diameter ranging from 5 down to 2 mm the radius of the diffraction pattern at the retina is 2 to 5 microns, in nice balance with the limitation set by the spacing of the receptors.

Of the various defects in image formation by the normal eye, the most serious is chromatic aberration. Because the index of refraction of the aqueous humor increases with shorter wavelength, the focus for blue light is nearer the cornea than that for red light. Leaving diffraction aside, a point object illuminated with white light can be translated at best into a "circle of least confusion" several microns in diameter (see Fig. 3-11). The effect of chromatic aberration would be worse were it not for the fact that the violet part of the spectrum is eliminated, both by absorption in the lens and by the poor sensitivity of the major cone pigments to violet light. As it is, the limitations imposed by aberrations are in harmony with those of diffraction and receptor spacing.

This harmony results partly from the choice of pupillary diameter. A larger pupil would reduce the effect of diffraction but would allow a larger and less "optically nice" part of the cornea-

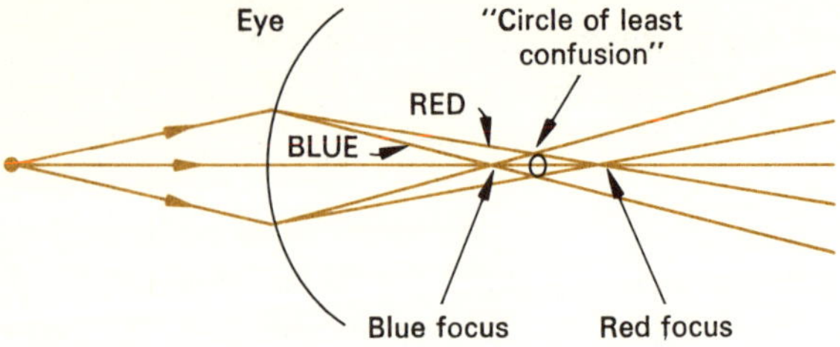

Fig. 3-11. Blue light is deviated more strongly than red at the surface of the eye. A point source of white light forms a succession of colored images along the optic axis; at best there is a circle of least confusion several microns in diameter.

lens system to take part in image formation. The hawk achieves an acuity of about one micron at the retina, apparently by using closer packing of the receptors, a larger iris, and an imaging system (cornea and lens) that is more nearly perfect over a large area.

As a final embellishment, the epithelium of the human fovea contains a bright yellow pigment that reduces the haze of scattered blue light. Because of this color the fovea is also called the *macula lutea* (yellow spot).

3-4. Absolute Sensitivity and Dark Adaptation

One quantum, absorbed by one rhodopsin molecule out of about 10^7, can cause the excitation of a rod (excitation means that the rod generates a signal that is communicated to other cells). This conclusion was drawn about 1940 by S. Hecht, S. Schlaer, and M. H. Pirenne from experiments on the threshold [1] of human vision (the least amount of light that can be seen). These investigators applied weak, short (about 1 msec) flashes of light to human subjects who were thoroughly dark-adapted. The flashes covered small areas of the retina containing about 300 rods. A threshold flash was defined as one having enough strength to be reported as seen

[1] A threshold stimulus of any kind is one that can barely be perceived. A higher threshold denotes a less sensitive system.

60 percent of the time (the subjects had been chosen for their reliability in not responding to dummy, or zero-energy, flashes). It was estimated that between 5 and 15 quanta absorbed in an area of 300 rods could cause a sensation of seeing with a probability of 60 percent. If 15 quanta are absorbed at random among 300 rods, one can compute the probability that a rod somewhere in the set receives more than one quantum.[1] This probability is only 30 percent, which allowed Hecht, Schlaer, and Pirenne to conclude that seeing can happen under such conditions that no single receptor cell receives more than one quantum. Later this conclusion was strengthened by experiments showing that the 5 to 15 quanta could bc distributed over about 2,000 rods and still give a 60 percent chance of seeing. Here the probability that more than one quantum is absorbed in a single rod is only 5 percent.

These experiments show that one quantum *can* excite a receptor cell but not that one absorbed quantum *will* do so. Actually the quantum efficiency for excitation, measured from the electrical responses of single receptors, is about 0.2 to 0.5 (0.2 in the squid according to W. A. Hagins).

Furthermore these experiments do not show that the excitation of one rod can cause a sensation of seeing. A cluster of 5 to 15 absorbed quanta can be seen, and the quantum efficiency for excitation is probably less than 0.5, so about two to seven excitations should suffice. Unfortunately this argument involves serious uncertainties in the estimation of quantum absorption and efficiency. The problem is best approached statistically,[2] in a way that circumvents these uncertainties.

First of all, it was found that a flash or cluster of quanta must be applied within a critical area (covering about 100 rods) and

[1] This problem is like the one of computing the chance that no two persons in a room have the same birthday, with 300 rods in place of 365 birthdays and quanta in place of persons. The probability that each quantum will find a different rod is $(299/300) \cdot (298/300) \cdots (286/300)$, or 0.70, so the probability that 2 or more quanta will fall on the same rod is $1 - 0.70$ or 0.30.

[2] The statistical analysis of threshold seeing was developed in the 1940s by several investigators including H. A. van der Velden, H. de Vries, M. A. Bouman, and E. Baumgardt.

a critical time (about 0.02 sec) in order to have its greatest effect. For flashes kept within this area and duration, the threshold energy is constant, but if either the area or duration is greater than critical, the threshold is greater. This suggests that seeing requires the coincidence of more than one excitation within the critical area and time. If a flash of a certain energy is reported as seen 60 percent of the time, the probability is 60 percent that the flash generates the required coincidence.

To confirm that a coincidence is needed and to determine the order of the coincidence (how many excitations must coincide), one must measure the frequency of seeing as a function of flash energy near threshold and compare this with computed probabilities for coincidences of various orders. Probabilities enter the problem because the number of quanta in a flash shows random variation, as does the number of raindrops hitting any one square of a sidewalk during a shower. Replicate flashes can be generated in a prescribed way, but the average number of quanta per flash is one thing, and the actual number in a given flash is another. Similarly there is randomness in the absorption of quanta and in the production of excitation by the rods.

In the experiments the flashes are kept within the critical area and time, so if a flash produces n excitations, it automatically generates an n-fold coincidence. Probabilities are computed from the Poisson formula: Let $\bar{n}$ be the average number of events per sample (raindrops per sidewalk square, or excitations per flash) and n be the actual number in a particular sample. The average $\bar{n}$ is obtained in principle from a very large number of samples. Then the probability that any particular value of n will occur, in one sample, is given by

$$P(n) = \frac{(\bar{n})^n e^{-\bar{n}}}{n!} \tag{3-2}$$

Thus if a flash of a certain strength produces two excitations on the average ($\bar{n} = 2$), the probability that it will actually produce *no* excitations is $2^0 e^{-2}/0!$, or $1/e^2$, since zero factorial (0!) equals

1. The probability that it will produce *one* excitation is $2^1e^{-2}/1!$ or $2/e^2$, and the probability of two excitations is $2^2e^{-2}/2!$, or again $2/e^2$.

Adding up the expressions [Eq. (3-2)] for $n = 0, 1, 2, \ldots, k - 1$ gives the probability that fewer than k excitations occur. One minus this sum gives the alternative probability, that k or more excitations will occur. Thus the probability of a k-fold (or more) coincidence is

$$P(n \geqslant k) = 1 - e^{-\bar{n}} \sum_{s=0}^{k-1} \frac{(\bar{n})^s}{S!} \tag{3-3}$$

The probability of a twofold (or more) coincidence is

$$P(n \geqslant 2) = 1 - e^{-\bar{n}}(1 + \bar{n}) \tag{3-4}$$

The way that $P(n \geqslant k)$ varies with $\bar{n}$ is sketched in Fig. 3-12, for k ranging from 1 to 6. The curve for $k = 1$ shows simply the probability that at least one excitation will occur, without regard for coincidences. Its shape is distinctly different from the shapes for $k = 2$ or more, where coincidences are involved.

The next step is to compare the data for frequency of seeing (as a function of flash energy) with these probability curves. This can be done without knowing actual values of $\bar{n}$ because the curves for different k values have distinctive shapes. In plotting the data on frequency of seeing, the horizontal axis in Fig. 3-12 can represent flash energy in arbitrary units instead of actual values of $\bar{n}$ (the two are proportional to each other). Now the trick is to adopt a horizontal scale such that the data will agree fairly well with one of the probability curves. It turns out that this can only be done in the range $k = 2$ to $k = 6$. For rod vision in humans, no expansion or compression of the scale will give a reasonable fit to the curve for $k = 1$, or for values of k greater than about 6. Therefore it could be concluded that seeing can result when fewer than 7, and perhaps as few as 2, rods are excited in a set of about 100 within 0.02 sec.

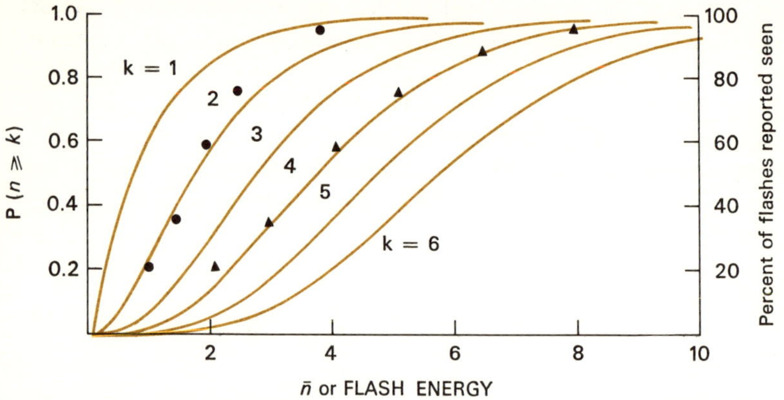

Fig. 3-12. The solid curves show the probability that if a flash produces n̄ excitations on the average, it will actually produce k or more excitations (see the text). The symbols ● and ▲ represent two attempts to fit a single set of data (frequency of seeing a flash, versus flash energy) to these curves. The flash energy is in arbitrary units, scaled differently for the two data-fitting attempts. The data are fairly compatible with all curves from k = 2 to k = 6, but not with the k = 1 curve.

The requirement of a coincidence of two or more excitations makes sense if an authentic stimulus is to be recognized against a backdrop of spontaneous excitation. Spontaneous activity is a general property of nerve systems; it comprises a background of "noise" that the brain should discount. One way that a system can filter out noise is to respond only to multievent coincidences.

Any mechanism for ignoring noise applies also to "real" signals, so a signal can be perceived reliably only if its strength exceeds that of the noise. Thus a lack of sensitivity might result from a high noise level as well as from a weak response to stimulation. With this in mind let us consider the phenomenon of dark adaptation.

The threshold for human vision varies over a wide range in response to illumination, reaching its lowest value (maximum sensitivity) after about 30 min in the dark. The progress of this dark adaptation after a light exposure is shown in Fig. 3-13. In this graph the ordinate is the threshold for seeing a spot of light that

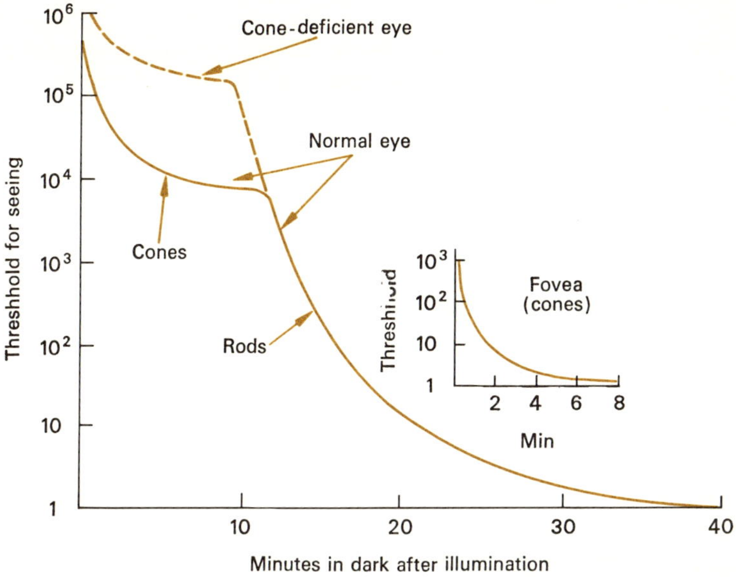

Fig. 3-13. Dark adaptation in human vision. The threshold declines in the dark after a period of illumination; maximum sensitivity is attained after 30 or 40 min. The two branches of the main curve reflect conditions in which the greatest sensitivity is provided by the cones and the rods respectively. Colors can be seen in the tests that define the cone-vision branch, but not under conditions of rod vision. The inset shows dark adaptation in the fovea. The scale of the ordinate in the inset should not be compared with that of the main figure, because the area of illumination in the foveal tests is much smaller.

covers much of the retina. The break in the curve shows the different properties of rods and cones. At longer times the rod vision, being far more sensitive than cone vision, determines the threshold. At progressively shorter times (less dark adaptation) the rods are less sensitive; at about 12 min postillumination the rods and cones contribute about equally. At still shorter times the rods are less sensitive than the cones, and the threshold is set by cone vision. In a cone-deficient eye the break in the curve comes earlier and higher (dashed curve in Fig. 3-13). The "cone" part of the curve for a normal eye shows approximately a fiftyfold change in sensitivity in about 12 min. If measurements are confined to the fovea, the cones there show a dark adaptation that spans a wider range in a shorter time: about 300-fold in 7 min as shown in the inset of Fig. 3-13. The sensitivity of rod vision increases about 3,000-fold after the break in the curve, but the cone-deficient eye reveals a greater range. In principle the range could be much greater, because the threshold for a nearly defunct eye would be very high.

During the photochemical activity that leads to excitation, the visual pigment goes through a cycle in which it becomes bleached and then restored (Sec. 3-6). For both rod and cone vision, W. A. H. Rushton has shown that the logarithm of the threshold is proportional to the fraction of visual pigment bleached. However, the proportionality is such that starting from complete dark adaptation, a 300-fold loss of sensitivity attends a bleaching of only 10 percent of the rhodopsin. Thus the loss of sensitivity cannot be explained by the absence of visual pigment, and it has seemed more natural (to Rushton and others) to attribute this loss to a much higher level of background activity or noise. On that basis a rod with 10 percent of its rhodopsin bleached is 300 times noisier than a dark-adapted rod. Since noise is equated to spontaneous excitation, it might be equated also to a spontaneous reaction of rhodopsin, induced by a thermal fluctuation rather than by light. One can then speculate that a rod containing some bleached rhodopsin is noisier, because the products of bleaching stimulate the thermally induced reaction. But this can be only a part of the mechanism of dark adaptation, as we shall see.

Figure 3-13 shows that for wide-area stimulation the sensitivity of rod vision is about 3,000 times greater than that of cone vision. Does this mean that cones are much "noisier" than rods, or are there other reasons?

The sensitivity of cone vision in the dark-adapted fovea approaches one-tenth that of rod vision in the periphery, provided that small test areas are used in the comparison. Most of this tenfold difference can probably be attributed to less dense pigmentation in the cones.

The statistical approach shows that for cones as well as for rods a few (perhaps only two) excitations must coincide within about 0.02 sec to give a sensation of seeing. However, the critical coincidence area in the periphery is one embracing about 100 rods, whereas in the fovea the critical area corresponds to only a few cones.

Putting these considerations together, the 3,000-fold difference between the threshold intensities for rod and cone vision (measured over a large area) can be explained partly by the weaker pigmentation of the cones but mainly by the greater convergence in the nerve pathways of the rods. Coincidences of excitation are amplified 100 times if a single ganglion cell receives signals from 100 rods.

In most animals the receptive field of one ganglion cell is typically much larger than 100 rods. It can be mapped (see Sec. 3-10) by recording the activity in one optic nerve fiber or one ganglion cell, while illuminating small groups of rods. In the cat the receptive field is often found to be as large as 10,000 rods, but many of these are engaged in mutually antagonistic effects. If illumination of one rod causes activity, the simultaneous illumination of another will suppress the activity. In general the rods in the center of the receptive field act in one way, reinforcing each other, and the surrounding ones antagonize them. Now, a curious aspect of dark adaptation is that the nerve function actually changes so that formerly antagonistic regions of the field become supportive. The center of the field, whose elements act in concert, grows and displaces the antagonistic ring around it. With exposure to light the antagonistic function is regained. The antagonistic activity is useful for emphasizing contrast, acuity, and sensitivity to movement, but the supportive activity gives the greater sensitivity needed in dim light.

We do not know how this change in nerve function is brought about.

Finally there is the rapid adaptation to different light levels through changes in the diameter of the iris. The diameter can change about fivefold, so the area (and consequently the relative amount of light admitted to the retina) can be changed by a factor of 25.

THE VISUAL PIGMENTS: THEIR CHEMISTRY AND FUNCTION

3-5. The Visual Pigments

In a dark-adapted eye exposed to weak light the most effective wavelength for human vision is 500 nm, corresponding to the absorption maximum of rhodopsin. In strong light the peak of the visibility curve (a plot of the reciprocal of the threshold versus wavelength; that is, an action spectrum for seeing) shifts to about 550 nm, reflecting the change from rod vision to cone vision. The maximum at 550 nm represents a composite of the cone pigments.

The agreement between visibility spectra and the absorption spectra of pigments provides the best proof that those pigments are responsible for vision. This proof is shown in greater detail, for rod vision, in Fig. 3-14: In the visible region the visibility curve coincides with an absorption spectrum of human rhodopsin and also with an action spectrum for the bleaching of rhodopsin. The differences in the ultraviolet, where there is absorption but no visibility, are explained in that the lens absorbs and thereby intercepts light of these wavelengths.

The foundations of our understanding of the chemistry of the visual pigments were laid in the 1870s by W. Kühne, who showed among other things that the visual pigment in isolated frog retinas was bleached by light and recovered in darkness. For our present knowledge of the chemistry of vision we are indebted mainly to the work of G. Wald and his associates, especially R. Hubbard.

The chromophores of visual pigments are derivatives of vitamin A (also called *retinol*—not to be confused with the aldehyde retinal, or retinene, discussed in Sec. 3-2 and below) in which the alcohol residue of the end carbon atom has been oxidized to an aldehyde group. This oxidation, shown in Fig. 3-15, is coupled to the reduction of NAD^+ to NADH in the receptor cell. The alde-

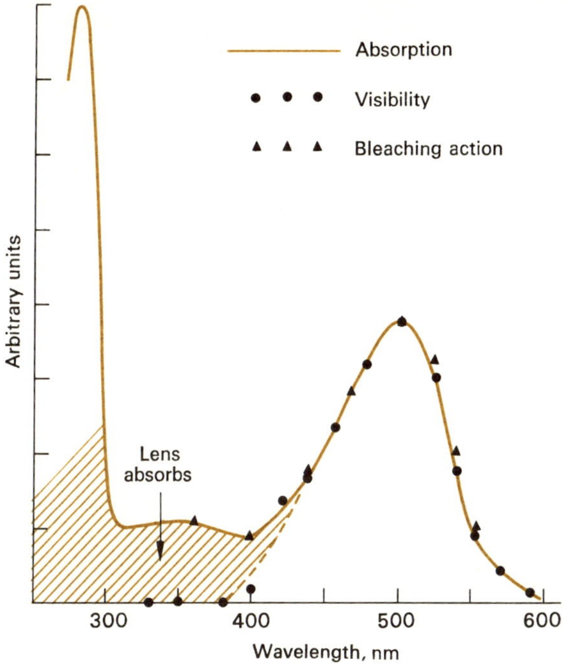

Fig. 3-14. The solid curve shows an absorption spectrum of a solution of rhodopsin (the 280-nm peak is due to the protein). The circles show the reciprocal of the threshold for seeing in dim light with a dark-adapted eye; this is an action spectrum for rod vision. The triangles show the effectiveness of light for bleaching rhodopsin in solution. The action spectrum for vision deviates from the other curves in the ultraviolet because the lens absorbs these wavelengths. Aside from that, the agreement between the data shows that rhodopsin is used for seeing.

hyde is called *retinal* (an unfortunate term, since the word *retinal* is also used as an adjective; see also Sec. 3-2). In nature it is found in two forms, corresponding to two forms of vitamin A. Figure 3-15 shows the most common form, retinal$_1$ (or R_1). The other form, retinal$_2$ (R_2), is shown in Fig. 3-16, together with the convention for the numbering of carbon atoms in these molecules. R_2 differs

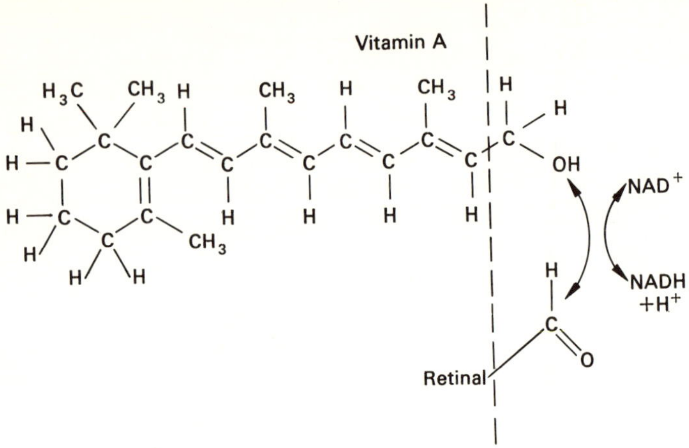

Fig. 3-15. Structures and interconversion of vitamin A and retinal, the chromophore of rhodopsin.

from R_1, and vitamin A_2 from vitamin A_1, in having a double bond between atoms 3 and 4. The longer chain of conjugated double bonds gives R_2 a longer-wave absorption maximum than R_1.

By combining R_1 or R_2 with the various opsins (proteins) found in rods and cones it is possible to generate a variety of visual pigments differing in the wavelengths of maximum absorption.

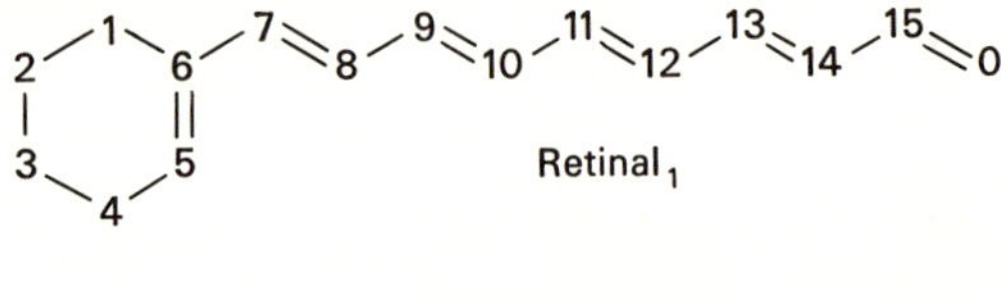

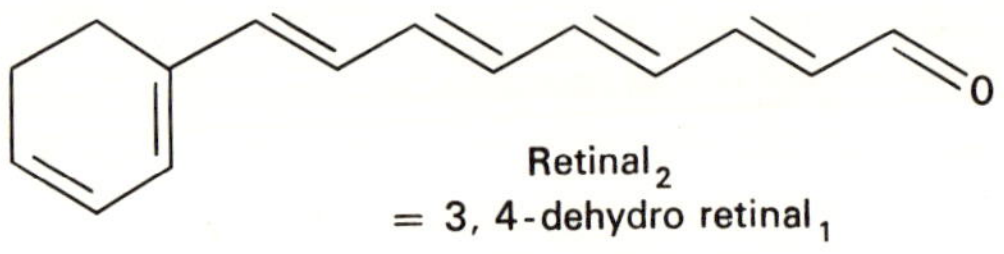

Fig. 3-16. Two types of retinal; correspondingly there are two types of vitamin A.

The ones found most commonly in nature are tabulated below, together with the scarcer cyanopsin. The different cone pigments responsible for color vision in man have not been isolated.

Table 3-1. Four Types of Visual Pigment. The Wavelengths of Maximum Absorption Vary Considerably from One Species of Animal to Another; the Most Common Values Are Shown.

$Retinal_1$ + rod opsin	Rhodopsin (most vertebrates)	(500 nm)
$Retinal_2$ + rod opsin	Porphyropsin (many fish)	(520 nm)
$Retinal_1$ + cone opsin	Iodopsin (most vertebrates having cones)	(560 nm)
$Retinal_2$ + cone opsin	Cyanopsin (tadpole but not frog)	(620 nm)

The retinals can exist in a variety of stereoisomers. The forms that are important in vision are the all-trans form, shown in Figs. 3-15 to 3-17, and the 11-cis form, shown in Fig. 3-17 along with the 9-cis isomer. These forms are interconvertible by a photochemical reaction: excitation of any one of them will produce a mixture of isomers, because the energy of excitation is enough to surmount the energy barrier (activation energy) between one form and another. On returning to the ground state the molecule will find itself in one isomeric form or another.

Both 11-cis and 9-cis retinal, but not the all-trans isomer, combine spontaneously with opsin. The functional visual pigment is the 11-cis form attached to opsin; we conclude this because if the protein in rods is denatured, the retinal can be extracted as the 11-cis isomer. The 9-cis form arises naturally in small amounts through photoisomerization and acts like an inert contaminant. Eventually it becomes restored to the active 11-cis form by another photoisomerization. The 11-cis isomer is converted to all-trans during the photochemical cycle of vision (see the next section). It is converted back to 11-cis both by light and by an isomerase enzyme. The 11-cis isomer is actually one of the less stable ones, because the methyl group on carbon 13 gets in the way of the hydrogen atom on carbon 10.

The bond between retinal and opsin is by a Schiff base link-

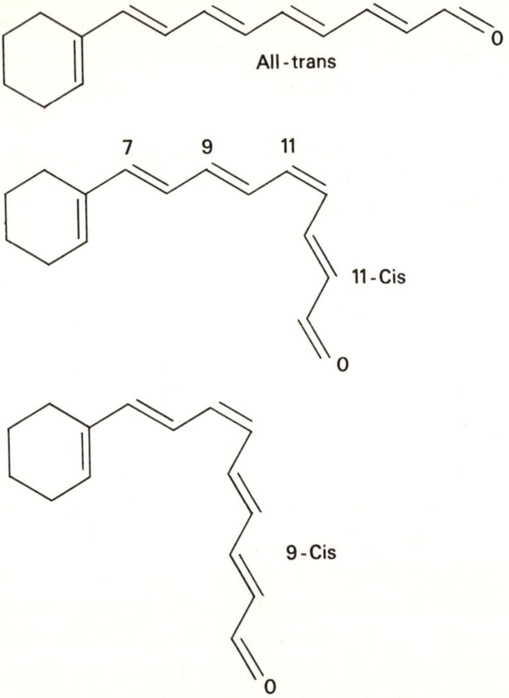

Fig. 3-17. Three stereoisomers of retinal. The 11-cis isomer is converted to all-trans in the photochemistry of vision.

age involving the aldehyde group at the end of the retinal molecule and the amino group of a lysine residue in the protein:

$$(\text{Ret})\text{–}\overset{\text{H}}{\text{C}}\text{=O} + \text{H}_2\text{N–Lys (opsin)} \rightarrow \text{R–}\overset{\text{H}}{\text{C}}\text{=N–Lys (opsin)} + \text{H}_2\text{O}$$

Recent experiments have indicated that retinal can form a Schiff base linkage with one of the lipids of the receptor cell membrane, phosphatidyl ethanolamine. These experiments have led to speculations that the chromophore actually moves between protein and lipid during the photochemical cycle. It would be premature to evaluate this suggestion.

3-6. The Chemistry of Vision

Light causes the bleaching of rhodopsin, in the intact eye, the excised retina, or a solution made by extracting the pigment from retinas with a detergent. If the retina is intact and metabolically active, the red color returns in the dark.

The bleaching is attended by the appearance of free all-trans retinal, derived from the 11-cis form bound to opsin. Recovery in the dark follows a roundabout path: The all-trans retinal is reduced to vitamin A, which diffuses from the receptor cell into the pigment epithelium. The vitamin A is converted enzymically to the 11-cis isomer, returned to the rod outer segment, and oxidized back to retinal. It then combines with opsin to form rhodopsin once more.

As an alternative to this complicated maneuver, if the all-trans retinal should absorb a quantum of light, it might be photoisomerized back to the 11-cis form and then recombine with opsin. This photorestoration probably plays a significant role in the eye of the squid, where the retinal never severs its link to the opsin.

Between the absorption of light by rhodopsin and the appearance of free all-trans retinal there is a succession of about five events, each one signaled by a characteristic change in the absorption spectrum. It is most reasonable to suppose that the very first event is the photoisomerization of 11-cis to all-trans retinal. The shape of the retinal molecule is then incompatible with the protein structure. The remaining events are plausibly regarded as upheavals in the conformation of the protein in response to this incompatibility. Finally the retinal comes off. In the 11-cis form it goes back on, forming the Schiff base link, and the protein returns to the configuration characteristic of rhodopsin. These events are shown somewhat pictorially in Fig. 3-18, along with the restoring events that complete the cycle.

Except for the first, the successive events after light absorption depend on the temperature, and this has provided a basis for their detailed study by T. Yoshizawa, R. Hubbard, R. Matthews, and others. In a typical experiment a solution of rhodopsin is brought to the temperature of liquid nitrogen, —196°C, and its

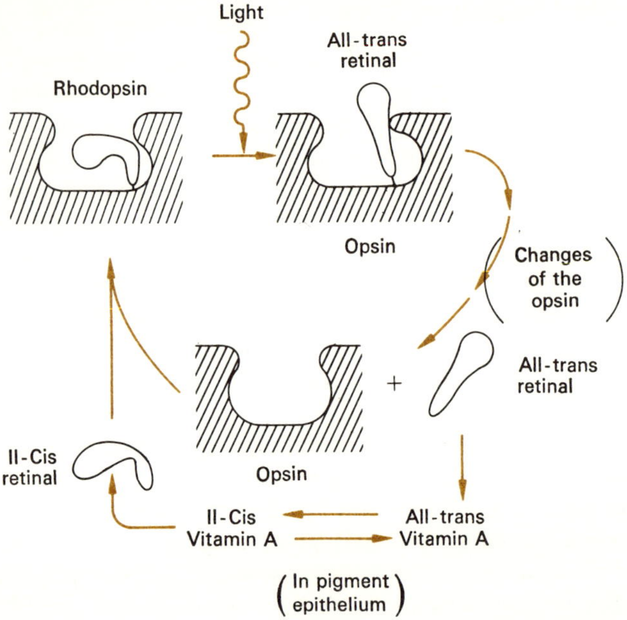

Fig. 3-18. A picture of probable events in the chemical cycle of vision. Light isomerizes 11-cis retinal to all-trans retinal, which is not compatible with the shape of the protein (opsin). These events somehow cause excitation of the receptor cell.

absorption spectrum is measured. A flash of light is applied, and a new measurement of the absorption spectrum shows a change: Some of the rhodopsin (absorption maximum at 498 nm) has disappeared, and a substance with an absorption maximum at 543 nm has been formed. This is the product of the first (photochemical) step in the sequence: all-trans retinal bound to opsin, called *prelumirhodopsin.*[1] Warming the sample now allows us to see all the subsequent steps in sequence, as summarized in Fig. 3-19. At a

[1] Originally a later product was thought to be the first and was named *lumirhodopsin.* When the "real" first product was discovered, it had to be called *prelumirhodopsin.*

temperature greater than about —140°C the 543-nm absorption band of prelumirhodopsin gives way to the 497-nm band of lumirhodopsin. Above —40°C [1] this changes to metarhodopsin I (487 nm), then above —15°C an equilibration between metarhodopsin I and metarhodopsin II is observed:

$$\text{Metarhodopsin I} \rightleftarrows \text{metarhodopsin II} + \text{H}^+$$

Metarhodopsin II is the first obviously bleached form; its absorption peak is at 380 nm, and it absorbs little in the visible. It enters an equilibrium with a new colored form, pararhodopsin (465 nm), above 0°C, and these substances finally yield a mixture of free all-trans retinal and opsin. The free all-trans retinal absorbs maximally at 387 nm. There is disagreement as to whether free retinal comes directly from metarhodopsin II, from parahodopsin, or from both.

At every step in the sequence the absorption of a quantum can cause the retinal to assume the 11-cis form and thus to give rise to rhodopsin again. These photochemical back-reactions are shown by the partly wavy arrows in Fig. 3-19.

The reactions of Fig. 3-19 can be made to illustrate nicely the phenomenon of the *photostationary state,* which will arise in an important way in Chap. 4. First consider a solution of rhodopsin at liquid nitrogen temperature, exposed to a very brief flash of light so that many of the rhodopsin molecules are excited. Some of the excited molecules will fall back into the 11-cis configuration (rhodopsin), and others will go over into the all-trans form (prelumirhodopsin). For simplicity of discussion we will ignore the fact that other isomers such as 9-cis are also formed. The result of the flash, then, is that some of the rhodopsin has been converted to prelumirhodopsin; the quantum efficiency of the conversion is given by that fraction of the excited rhodopsin which becomes prelumirhodopsin. These events are shown in Fig. 3-20*a*. Now that

[1] The transitional temperatures are not sharply defined; the statement that lumirhodopsin changes to metarhodopsin I above —40°C means only that the reaction is very slow below this temperature.

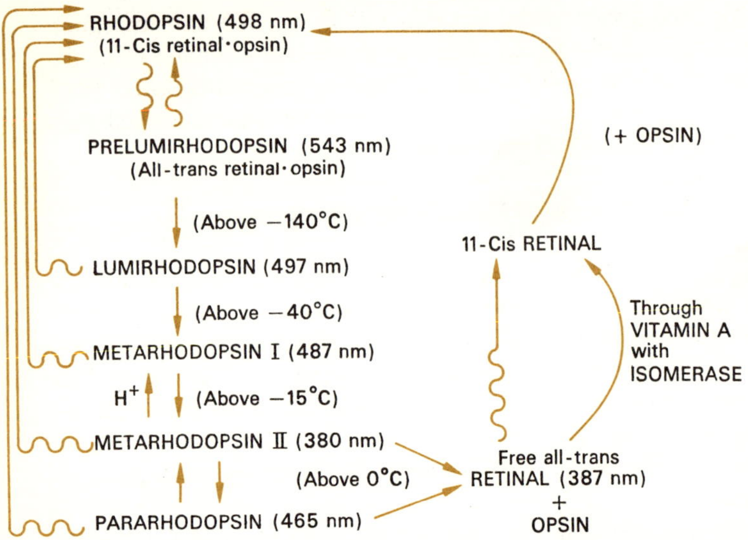

Fig. 3-19. Details of the chemical cycle of the visual pigment in vertebrates, together with photochemical back-reactions. All light-driven steps are shown by wavy arrows. The designations of temperature show roughly where the reaction rates become appreciable. The wavelengths indicate the absorption maxima.

some molecules of prelumirhodopsin have been formed, a second flash can excite them as well as the rhodopsin molecules. The excited molecules of both kinds can then decay partly to one isomer and partly to the other. Under steady illumination the system approaches an equilibrium, the photostationary state, in which rhodopsin and prelumirhodopsin are being interconverted at equal rates. The relative proportions of the two pigments in the photostationary state are determined by two factors: The first is the relative rates at which the pigments absorb light so as to become excited; this depends on their concentrations and on their extinction coefficients at the wavelength(s) of excitation. The second factor is the partition of photochemical products: for each kind of excited molecule, the relative probability that it will decay into rhodopsin or prelumirhodopsin (in the present case these probabilities are

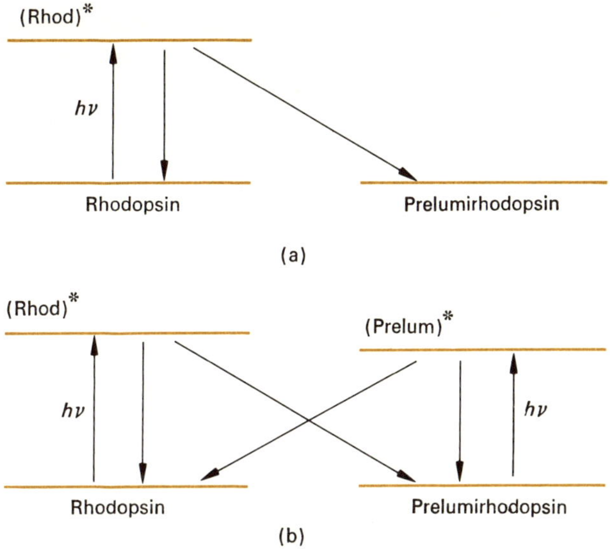

Fig. 3-20. (a) If a solution of rhodopsin at liquid nitrogen temperature is exposed to a very brief flash of light, some of it is converted to prelumirhodopsin. (b) Under sustained illumination the two pigments are interconverted in a steady cycle of excitation and decay. The relative proportions of the two pigments in this photostationary state are determined by the individual reaction rates.

about equal). The interplay of these reactions is shown in Fig. 3-20*b*.

If the two substances have different absorption spectra, as they do in this case, the photostationary state can be shifted to favor one substance or the other by a suitable choice of exciting wavelength. Light of wavelength 600 nm is absorbed far more by prelumirhodopsin than by rhodopsin, so that this wavelength will drive the system almost completely toward rhodopsin. Conversely the use of 450-nm light, absorbed more strongly by rhodopsin, will cause prelumirhodopsin to predominate in the mixture.

A photostationary state can also arise when the interconversion of two or more pigments is less direct. In the squid retina the excitation of rhodopsin leads to the formation of metarhodopsin A, a pigment analogous to the vertebrate metarhodopsin I. Under physiological conditions the process goes no further, and metarhodopsin A is converted back to rhodopsin photochemically. During illumination the rhodopsin and metarhodopsin A exist as components of a photostationary state, and as opposite poles of a cycle

$$\text{Rhodopsin} \rightsquigarrow \longrightarrow \longrightarrow \text{metarhodopsin A}$$
$$\nwarrow\!\!\underline{\hspace{12em}}\!\!\sim\!\sim$$

that causes visual excitation.

3-7. The Fast Photovoltage

It has been known for decades that if a sensitive electrode is placed on the cornea and another behind the retina or in the mouth, light entering the eye will cause changes in the voltage measured between the electrodes. A record of this effect is called an electroretinogram (ERG; see Fig. 3-21). The ERG reflects electrical activities in the receptors and other nerve cells during excitation. It makes its appearance a few milliseconds after the start of illumination or after a brief flash.

In 1964 K. T. Brown and M. Murakami, working with monkeys, reported another electrical effect, measured in the same way but using flashes far stronger than are needed to give a conspicuous ERG. The new effect begins with no detectable lag (less than 1 μsec) and is completed within a few milliseconds. It thus occupies the time between the illumination of rhodopsin and the onset of the ERG. This rapid effect is usually called the *fast photovoltage* (FPV), although it was originally named the *early receptor potential* (ERP). It was missed for years because of the strong light needed to evoke it.

In contrast to the ERG, the FPV is immune to the action of poisons and antiphysiological conditions (such as the replacement of sodium in the surroundings by lithium) unless these conditions disrupt the organized structure of the retina. The FPV is seen in a

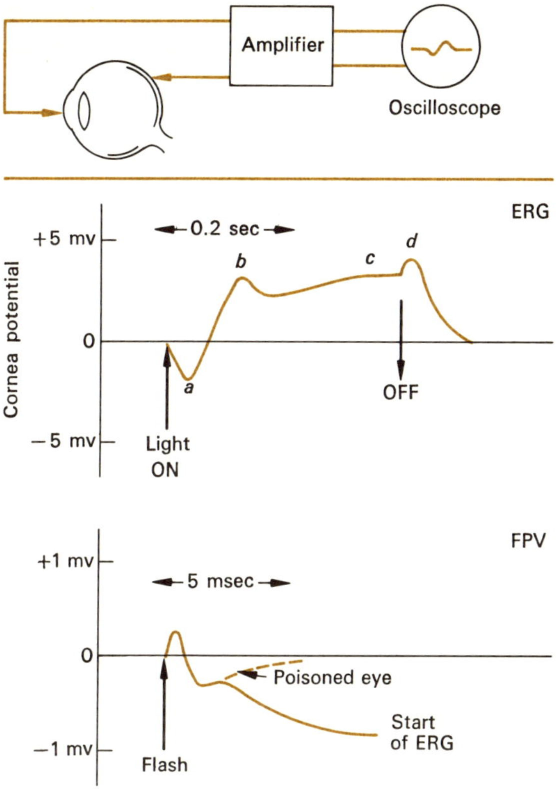

Fig. 3-21. Electrical effects in the vertebrate eye: the electroretinogram (ERG) and the fast photovoltage (FPV). The latter requires much stronger light than the former to be seen.

metabolically inert retina, as long as the bleaching of visual pigments can occur, but not in a solution of rhodopsin.

Extensive recent studies by R. A. Cone, W. L. Pak, E. B. Goldstein, T. G. Ebrey, and others have shown that the FPV is an electrical change that accompanies certain steps in the chemical cycle of vision. It is useful, therefore, as an adjunct to the optical study of these reactions.

The size of the FPV is proportional to the amount of visual pigment bleached, up to a maximum of about 0.5 mv. In rat's eyes the action spectrum for its generation matches the absorption spectrum of rhodopsin, but in frog's eyes it is associated with the cone pigment. Probably the electrical effect is produced in the rods as well, and also in a solution of rhodopsin, but to see it as an FPV the responsible molecules must all be lined up in one direction. This would be expected if the effect were due to electric dipoles on a molecular scale. If so, the size of the FPV corresponds to a dipole of one electron and one equal positive charge, separated by 1 Å, for each molecule of pigment bleached.

As shown in Fig. 3-21, the FPV has two phases: a positive wave, called R_1, followed by a negative wave called R_2. Below about $-15°C$, R_2 disappears leaving only R_1. The rise of R_1 is associated with the formation of metarhodopsin I, and R_2 is associated with the conversion of metarhodopsin I to metarhodopsin II.

Considerably more can be learned by observing the FPVs in response to closely spaced pairs of flashes. The first flash starts the chemical cycle and is attended by the usual FPV with phases R_1 and R_2. A second flash can catch some of the intermediates of the cycle, driving them photochemically back to rhodopsin. These reactions are again attended by characteristic phases in the FPV. Thus the return of metarhodopsin I to rhodopsin gives a negative signal called R_A. It is like an inverted R_1 signal, and the action spectrum for its production corresponds to the absorption spectrum of metarhodopsin I. The return of metarhodopsin II to rhodopsin gives a positive signal (the opposite of the R_2 wave) called R_B, and a third signal, R_C, can be identified with the return of pararhodopsin to rhodopsin.

These electrical effects are summarized in Fig. 3-22. Their origin might be the formation of electric dipoles attending the chemical steps; this could happen if electrically charged regions of the protein become exposed as a result of the reaction. The rise of the voltage wave then reflects the progress of the reaction, and the decay and disappearance of the signal could take place be-

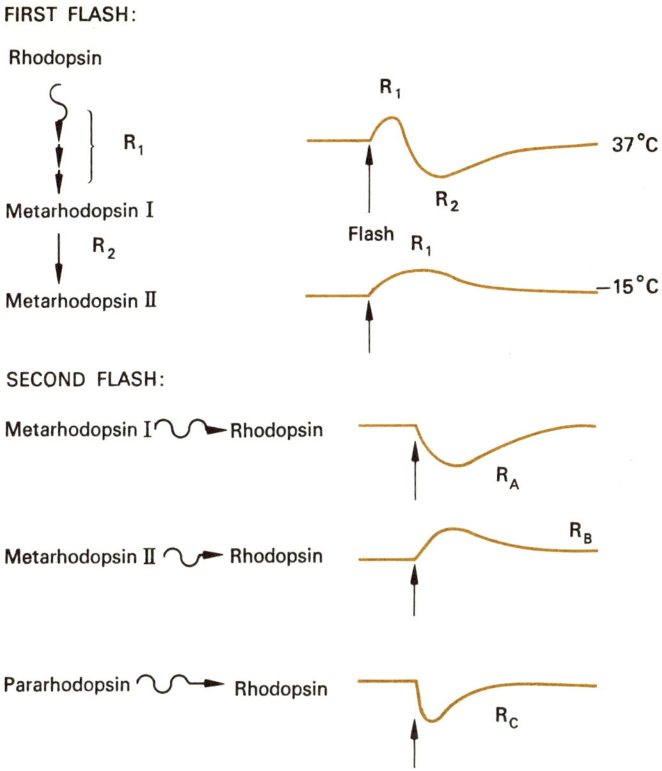

Fig. 3-22. Components of the FPV and the chemical steps with which they are associated. See the text for a discussion.

cause the dipoles are neutralized by the movement of ions. Each new chemical change disturbs the neutralized charge pattern and can give a transitory FPV.

After one flash has set the chemical cycle in motion, the second flash reveals the amounts of some of the intermediates: the amount of metarhodopsin I is measured by the size of the FPV component R_A; the amount of metarhodopsin II, by R_B; and the amount of pararhodopsin, by R_C. By varying the time at which the second flash is applied one can deduce the kinetics of formation and disappearance of these intermediates. The results for the rat are approximately as follows:

1. Rhodopsin ⟶ prelumirhodopsin ⟶ lumirhodopsin, less than 1 μsec (0°C)

2. Lumirhodopsin ⟶ metarhodopsin I, less than 1 msec (0°C)

3. Metarhodopsin I ⟶ metarhodopsin II, about 1 sec at 0°C, 40 μsec at 37°C

4. Metarhodopsin II ⟶ pararhodopsin, about 2 min at 37°C (body temperature)

5. Free retinal from metarhodopsin II and/or pararhodopsin in about 20 min at 37°C

The electroretinogram shows that excitation, as manifested by changes in the membrane potentials of nerve cells, does not begin until about 3 msec after a flash. The reaction rates deduced from the FPV then show that excitation begins before the chemical cycle has progressed appreciably beyond metarhodopsin II. Recall that in the squid the cycle which causes excitation does not seem to go beyond metarhodopsin A, the counterpart of metarhodopsin I.

We do not know whether the electrical change manifested by the FPV is an essential part of the process of excitation or whether it is merely an incidental attribute of the chemistry.

3-8. Color Vision

Two wavelengths of light can be distinguished if they cause different relative stimulation of two or more types of receptor cells. This can lead to different patterns of activity among the various cells, and the brain can interpret these patterns in terms of color.

Although two kinds of receptor with different absorption spectra are enough to provide some color sense, it was proposed by Thomas Young and L. von Helmholtz about 1800 that there are three basic types. In 1860 the physicist James Clerk Maxwell showed that nearly all colors can be reproduced by mixing three basic colors in suitable proportions. It does not matter how the three primary colors are chosen as long as they are "different enough," that is, if no one of the three can be made by mixing the other two. Yellow, blue, and green would not do, but yellow, red, and green would. Since the choice of primary colors is somewhat arbitrary, we cannot learn much about the spectral properties of the types of receptor by doing color-mixing experiments.

Something can be learned about the receptors by studying the ability of the eye to discriminate changes of color along the spectrum. The sensitivity to a change of wavelength should be large if there is a large change in the relative stimulation of two kinds of receptor. This condition will prevail where the absorption spectra of two types of receptor overlap, as shown in Fig. 3-23. If we plot (stimulation of type A)/(stimulation of type B) against wavelength, the steepest part of this curve is where the relative stimulation changes most rapidly. This should represent the region of greatest sensitivity to a change in the wavelength. Working backward, one can hope to deduce the absorption spectra of the receptors from data on the sensitivity to color contrast.

A glance at an actual spectrum [1] will bring home the fact that the colors seem to change most rapidly in the green-yellow-orange region, about 550 to 600 nm.

The analysis of color discrimination was exploited most carefully by W. S. Stiles in the 1940s and 1950s. He concluded that human color vision is based on three kinds of receptor: major ones with absorption maxima at about 540 and 580 nm, and a lesser one centered at 440 nm. His predictions have been borne out fully by more recent direct measurements of the optical properties of cone pigments.

Methods of the direct observation of the pigments and their bleaching in the human eye have been developed in the past decade by W. A. H. Rushton and by J. D. MacNichol. The technique is one of reflection spectrophotometry. A narrow beam of monochromatic light is sent into the eye, and a small fraction of this light emerges after reflection from the pigment epithelium. The

[1] To see spectral colors make a water prism by holding a mirror at an angle under the surface of a basin of water.

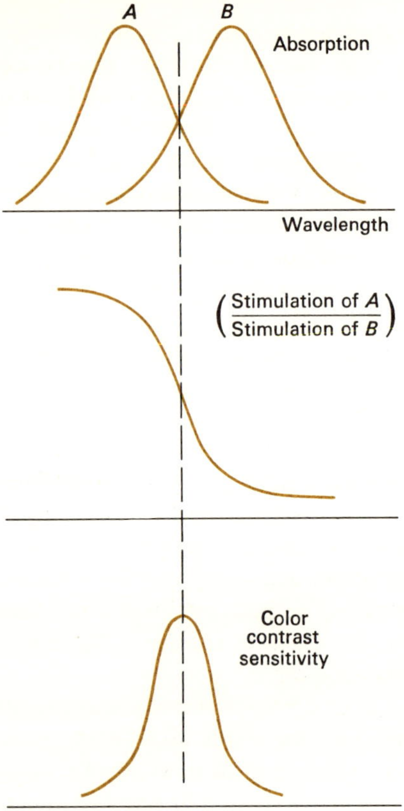

Fig. 3-23. Color discrimination can result from difference in the relative stimulation of two kinds of visual pigment. This drawing shows hypothetical pigments A and B (top) and a plot of the way that the relative stimulation varies with wavelength over the region encompassed by the absorption bands (center). In the steepest part of this plot, where the relative stimulation changes most rapidly with wavelength, the sensitivity to a change in wavelength should be greatest (bottom).

emerging light has passed twice through the layer of receptor cells. While this measurement is going on, a stronger flash can be applied so as to cause bleaching of the visual pigments. The change in optical density in the receptor cells is observed as a change in the intensity of the reflected measuring beam. The technique is very difficult, because the black pigment epithelium reflects only a small fraction of the entering light.

Rushton first directed the measuring beam to the rods and observed the well-known bleaching of rhodopsin, to show that his method was reliable. He then confined the beam to the fovea so as to study cone pigments and made the problem simpler by using subjects with defective color vision. These persons lacked one or another type of cone pigment, so the confusion of overlapping spectra was minimized.

One class of color-blind subject, the protanope, was found to have a single predominant cone pigment absorbing maximally at 540 nm. In another kind of color-blind person, the deuteranope, this pigment was missing, and a cone pigment with maximum absorption around 570 or 580 nm was found. In persons with normal vision both pigments could be recognized. Rushton called these cone pigments *chorolabe* (540 nm) and *erythrolabe* (580 nm).

A comparison with Stiles' analysis would suggest that a third kind of cone pigment should be sought: a less abundant pigment with an absorption maximum at 440 nm. Direct evidence for this "cyanolabe" has been only marginally convincing because of technical difficulties encountered in the blue-violet part of the spectrum.

The human visibility curve for seeing in bright light is effectively an action spectrum for cone vision. It is usually drawn as a smooth curve with a single maximum at about 550 nm, but the most careful measurements show that it has a two-humped shape with a suggestion of maxima at about 540 and 570 nm.

The original proposals of Young and Helmholtz thus seem to be borne out by modern investigations, but with indications that most of the burden of color discrimination is carried by the two types of cone pigment that function best in the green-yellow-red regions.

Another way to identify distinct color sensors in the eye is to measure electrical responses of single receptor cells (cones) to illumination at different wavelengths. The responses are changes in electric potential across the cell membrane; the significance of these changes and the technique of their measurement will be described in the next section. For the present let us accept that these electrical responses reflect excitation of the receptor cells. A. Kaneko, making such measurements with the carp, has found three kinds of cone: one responding maximally to blue light (peak at 450 nm), one to green (peak at 520 nm), and one to yellow (620-nm peak). Similar results are being obtained by others with eyes of other vertebrates.

Actually the richness of the color sense is enhanced greatly if the visual field has abundant details of form and contrast. This was shown dramatically by Edwin Land in certain demonstrations of "two-color" projection. Land photographed a single scene (a landscape, or a collection of colored objects) in duplicate. The two exposures, *on black-and-white film,* were identical except that a red filter had been placed in front of the lens for one of the shots. The resulting black-and-white lantern slides differed only in the degrees of shading of different areas. The two slides were then projected so that their images could be superimposed on a screen. The picture that had been photographed through a red filter was projected through the same red filter. If only this "red" slide was projected, the screen showed a picture in various shades of black and red, as would be expected. The other slide by itself of course produced an ordinary black-and-white image. But when the two were superimposed, the scene appeared in approximately its original colors!

No convincing explanation has been offered for this remarkable result. The phenomenon is striking only if the patterns of light and dark are complicated, as they would be in a natural landscape. Evidently the interpretation of color by the brain involves an intricate interplay in the relative stimulations of various cones. Complex patterns in time as well as in space are important; an impression of color can be gained by looking at a spinning top painted with suitable patterns of black stripes on a white background (Benham's disk).

To see how a sense of color could be induced with a suitable pattern of white light, imagine that a screen is contrived so that light coming through it will cast spots only on the "red" receptors in the retina. The brain will interpret the pattern of excitation as having been caused by red light.

The basis of color discrimination does not need to be provided by different types of visual pigment. In the chicken retina there is just one kind of cone pigment, iodopsin, but there are yellow-, green-, and red-colored oil droplets in front of the cones, producing three spectrally distinct classes of receptor.

FROM QUANTA TO NERVE IMPULSES: THE TRANSDUCTION AND PROCESSING OF VISUAL INFORMATION

3-9. Transduction

A nerve cell ordinarily maintains an electric potential across the membrane, such that the outside is more positive than the inside. If this state of polarization is nullified in a small region of the membrane, as by a local outward current, a self-amplifying wave of depolarization may develop and spread to adjacent parts of the membrane. This wave is then propagated along the nerve cell or fiber, carried by currents of ions passing through the membrane. Behind the advancing wave the membrane recovers its resting state of polarization, so the wave moves along the nerve fiber as a "spike" of depolarization. This traveling wave is called a *nerve impulse,* or an *action potential.*

An action potential can be initiated within a nerve cell by chemical activities that cause local depolarization. This happens in the receptor cells of the squid eye. Alternatively an action potential can be initiated at a synapse, by a chemical agent transmitted from the adjacent cell. This is how the action potentials in vertebrate ganglion cells begin.

The cell that stimulates another across a synapse may itself carry action potentials, or it may only generate a slower change in its membrane potential, lacking the autocatalytic and traveling properties of the action potential. These slower "generator potentials" are commonly observed in the cells of sensory systems. They occur in visual receptor cells and also in the interneurons (horizontal, bipolar, and amacrine cells) of the vertebrate retina. They are the source of the various waves of the ERG as shown in Fig. 3-21.

The generator potential in a receptor cell is graded: its magnitude varies with the intensity of stimulation. The action potentials are all-or-none; their magnitude is invariant in one nerve cell. The size of a generator potential, which reflects the intensity of stimulation, is translated into a greater or lesser frequency of action potentials. Usually a nerve cell fires a continuous volley of action potentials, and the frequency of these is modulated under the influence of a generator potential. In the squid eye these events all take place in a single receptor cell: light causes a graded change in the polarization of the membrane, and this generator potential modulates the frequency of a sequence of spontaneous action potentials. These are carried in optic nerve fibers which are outgrowths of the receptor cells.

In the lateral eye of the horseshoe crab the receptor cells (called *retinula cells*) develop a depolarization in response to light, and this is followed by the appearance of action potentials in the adjacent optic nerve cells (called *eccentric cells*).[1]

The foregoing information has been derived for visual systems by the techniques, developed over the past half century by E. D. Adrian, R. Granit, and H. K. Hartline among others, for implanting microelectrodes in selected places so as to study the electrical activities of single cells or small groups of cells during stimulation. In recent years the electrical behavior of the squid retina has been explored with particular success by W. A. Hagins. With the verte-

[1] This picture has been accepted for years, but it now appears possible that the visual pigment and the origination of electrical responses to light are situated in the eccentric cells, at surfaces near the retinula cells. This would leave the function of the retinula cells in doubt.

brate retina the investigations of T. Tomita and J. E. Dowling have been especially successful. A helpful technique here has been to inject a dye down the axis of a capillary pipet electrode after its implantation, to locate the tip and show which cell has been impaled. The fluorescent dye Procion Yellow permeates the impaled cell body and all its outgrowths (fibers, dendrites), giving a beautiful visualization of the cell and its ramifications. In these studies the mud puppy has been a useful animal, because its visual cells are unusually large.

Hagins has shown that when light is applied to the pigmented part of the squid retina, a current of sodium ions is induced to flow into the receptor cells at the site of illumination. The Na^+ ions pass from the fluid surrounding the cells, through the cell membrane to the interior. This local inward current leads to a compensating outflow across other parts of the cell membrane as dictated by the ion conductance of the membrane. The outward current causes depolarization of the membrane at the nonpigmented end of the cell, where the optic nerve fiber originates. This is followed by an increased number of action potentials in the nerve fiber.

The work with invertebrates such as the horseshoe crab and the squid showed that light causes a depolarizing current, so it came as a surprise when Hagins showed that the reverse happens in vertebrate receptor cells. The vertebrate rod or cone maintains a loop of sodium ion current: inward in the outer segment where the pigment is and outward in the inner segment which communicates with other nerve cells. Light striking the outer segment causes a *decrease* in this current, making the membrane of the inner segment become more strongly polarized (outside-positive) than in its resting condition. This transient hyperpolarization is seen as the *a*-wave in the ERG (Fig. 3-21).

The horizontal cells also become hyperpolarized in response to the change in activity of the receptor cells. The bipolar cells show both hyperpolarization and depolarization in a way that will be mentioned in the next section. The amacrine cells respond by becoming depolarized and occasionally show action potentials. In the ganglion cells one sees slight depolarization and numerous action potentials.

The basic question in visual transduction is, how does one quantum initiate an electrical effect involving a large number of ions? How can an excited rhodopsin molecule anywhere in the receptor cell cause the development of a generator potential? Some insight into these questions has resulted from Hagins' work with the squid retina. Hagins found that the inward sodium current is localized where light strikes rhodopsin in the receptor cell membrane. Lithium cannot replace sodium in carrying the current. One quantum which starts the chemical cycle of rhodopsin can cause an influx of about 10^5 Na^+ ions. This pulse of current begins about 80 msec after the quantum has been absorbed and lasts about 100 msec. Before it appears, there is the rise and fall of an FPV associated with the formation of metarhodopsin A. For an influx of 10^5 ions the compensating efflux in a remote part of the cell can change the potential by about 20 μv, easily distinguishable from the noise level of about 1 μv which usually prevails.

The details in the vertebrate eye are different. The electrical effect amounts to about 10^5 ions per effective quantum, but it is a decrease in the inward current. The FPV is complete and the generator potential is evident within about 5 msec. In spite of these differences one can at least visualize a plausible sequence of events for transduction in general: The isomerization of 11-cis retinal to all-trans exposes an area of the membrane about 10 Å across. This opens a pore large enough (7 Å) to let Na^+ leak through, or closes a preexisting pore, or else exposes the active site of an enzyme that controls a sodium pump. The current (or change of current) ends when local supplies of the necessary chemicals are depleted or when the site is restored to its "dark" condition.

This model of transduction is not hard to visualize if the rhodopsin is part of the external cell membrane, as it is in the squid receptor and in the vertebrate cone. In the vertebrate rod we must deal somehow with the problem that the pigment is in the membrane of an interior thylakoid, distinct from the external membrane which supports the generator potential.

3-10. The Processing of Visual Information in the Retina

The eye makes continual tiny movements, so any point in the image is shifting in an irregular way from one receptor cell to another. This wandering can be prevented, either by narcotizing the eye muscles and clamping the head or by supporting the perceived object on a holder fastened to the eye itself. Under these conditions most objects seem to disappear. If an object is moved so that its image shifts on the retina, it reappears momentarily and then becomes invisible again. This fascinating effect, which shows the importance of movement for seeing, can be approximated by relaxing and staring fixedly at one point without blinking or moving. After a time a visual "whiteout" develops: the entire peripheral visual impression becomes lost in a sort of fog.

A related effect is that the contrast between a black field and a white one seems greatest at the boundary. Figure 3-24 shows this by means of a grid of black lines. If you look straight at one of the intersections, you may notice evanescent grayish blobs in the other intersections.

These effects indicate that if the input to a set of receptors is constant, the message to the brain fades or becomes ignored as time goes by. Stimulation must be turned off and renewed in order to be effective. *Visual fatigue* is manifested in many ways: If a motorist stops after driving steadily for a long time, the surroundings seem to move forward, giving the disquieting impression that the car is rolling backward. If a pair of tinted glasses is worn for some time, the color becomes forgotten and the colors of the landscape seem natural.

There is obvious survival value connected with the enhancement of contrast and with a special sensitivity to movement. We should like to understand how these advantageous effects come about, in terms of the activities and interactions of individual cells.

Fig. 3-24. If you stare at one of the intersections of black stripes, you may become aware of evanescent grayish blobs in the other intersections.

Most of our present knowledge has come from the measurement of electrical activities of single cells, using microelectrodes.

The work of the 1940s and 1950s (H. K. Hartline with horseshoe crabs and frogs, S. W. Kuffler with cats and molluscs, and R. Granit with insects, among others) involved mainly the recording of action potentials from optic nerve fibers during illumination of clusters of receptor cells. In general three patterns could be found in different fibers of a single optic nerve. These patterns, shown in Fig. 3-25, are called *on, off,* and *on-off* for reasons that are obvious from thc figure. The off response can be correlated with the *d*-wave of the ERG (Fig. 3-21).

To understand the off effect we must appreciate the consequences of reciprocal inhibition in the nerve paths. This can be illustrated with the eye of the horseshoe crab, which ordinarily shows just *on* responses. If the measuring electrode touches a single optic nerve cell, one can find which cluster of receptor cells

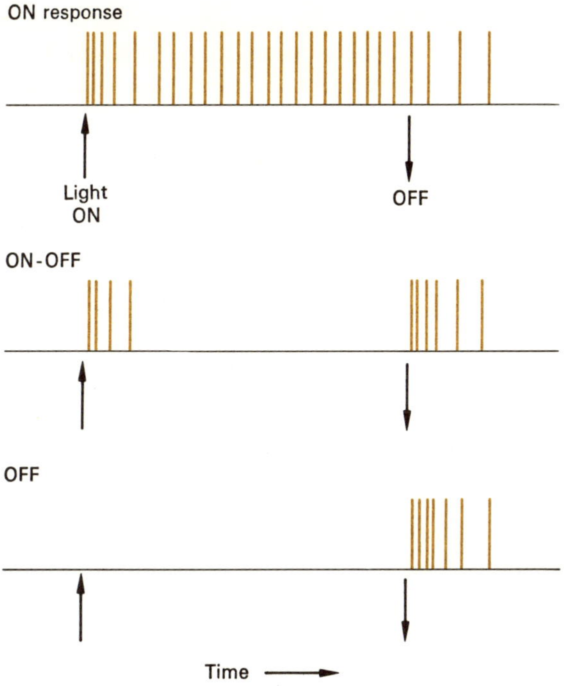

Fig. 3-25. Three kinds of pattern in the action potentials measured in an optic nerve fiber or ganglion cell, in response to illumination.

causes the greatest response by moving a tiny spot of light over the surface of the eye. Suppose that the most sensitive cluster has been found and the light spot is then enlarged so that more receptors are illuminated while the original cluster receives as much light as before. The result is a *decrease* in the frequency of action potentials in the nerve fiber. Illuminating the receptor cells around the central cluster inhibits the action of the central ones. These effects are reciprocal; with another nerve fiber the roles of central and inhibiting receptors are interchanged.

Once this reciprocal inhibition had been recognized, the pattern of illumination could be arranged so as to cause an *off* response, never before seen in the horseshoe crab. The on discharge from one group of receptors could be suppressed by sufficient illumination of the nearby inhibitory ones. Then when the light was turned off, the persistence of the on activity outlasted the inhibitory action. A burst of action potentials, coming after the end of illumination, reflected this release from inhibition.

The behavior of the horseshoe crab provides a model for understanding *off* effects, contrast enhancement, and sensitivity to movement in terms of an interplay of stimulating and inhibiting effects. The model is limited in two ways: First, it tells us nothing at a molecular level of understanding about how the antagonistic effects are produced. Second, the pattern of interactions in the vertebrate retina is much more complicated.

The neurological functioning of the vertebrate retina has been studied very actively in recent years, by J. W. Lettvin with the frog, H. B. Barlow with the cat, H. Maturana with the pigeon, T. Tomita with fish and squirrels, A. Koneko with carp and dogfish, J. E. Dowling with the mud puppy, and most extensively by D. H. Hubel and T. N. Wiesel with cats and monkeys. For the most part this work has been a mapping of input-output relationships: an electrode is placed on or in a ganglion cell (or a bipolar cell, etc.), and the receptive field serving that cell is delineated by stimulating smaller and larger sets of receptor cells. A list of some principal observations will show the nature of this work and will display the versatility of the retinal nerve cells in processing the input to the receptors.

Ganglion cells can be classified broadly as *contrast* or *movement* detectors. The former are for seeing details of form; they are predominant in cats and monkeys. The latter, which respond only to moving patterns, are especially abundant in pigeons and amphibians (in the mud puppy the ratio is about 4:1 in favor of movement detectors).

Dowling's work with the mud puppy indicates that the contrast ganglion cells are in direct contact with bipolar cells, but the movement ganglion cells are linked to amacrine cells which are in turn connected to bipolar cells (refer to Figs. 3-4 through 3-6).

The movement cells often respond only to movement in a particular direction. Any pattern of light and shadow moving across the receptive field causes a maximal burst of action potentials when the motion is in one direction but much less or none in all other directions.

The contrast cells come in an interesting variety. Some respond to a straight line lying on the receptive field in one direction but not in other directions. Some respond to the presence of any convex form: a closed figure such as a circle or a triangle, but not to a straight edge.

Most contrast cells respond simply to light falling on the central part of the receptive field, either with an *on* or an *off* response. Whichever of these responses is evoked by illuminating the center of the receptive field, it is suppressed when light strikes a surrounding region. This is sketched in Fig. 3-26, where the central or stimulating part of the receptive field is labeled *a* and the inhibitory part *b*.

This pattern of antagonism between central and peripheral regions of the receptive field is found in movement cells as well as in contrast cells.

Not only does the outer region *b* inhibit the discharge of action potentials in the ganglion cell, but it also produces an effect opposite to that of region *a* on a bipolar cell. An electrode in a bipolar cell shows generator potentials when parts of the receptive field are illuminated. Illumination of the center of the field evokes either hyperpolarization or depolarization in a bipolar cell; this probably bears a direct connection to *off* and *on* (or vice versa?) responses in a related ganglion cell. In either case, if outer region *b* is illuminated, the wave of hyperpolarization or depolarization is followed by a wave of the opposite polarity.

The inner region *a* agrees roughly with the spread of direct connections from one bipolar cell to receptor cells. Region *b* corresponds to the spread of connections from a bipolar cell to receptor cells by way of horizontal cells. Thus the inhibitory action seems to involve messages that reach a bipolar cell from a receptor cell through an intervening horizontal cell.

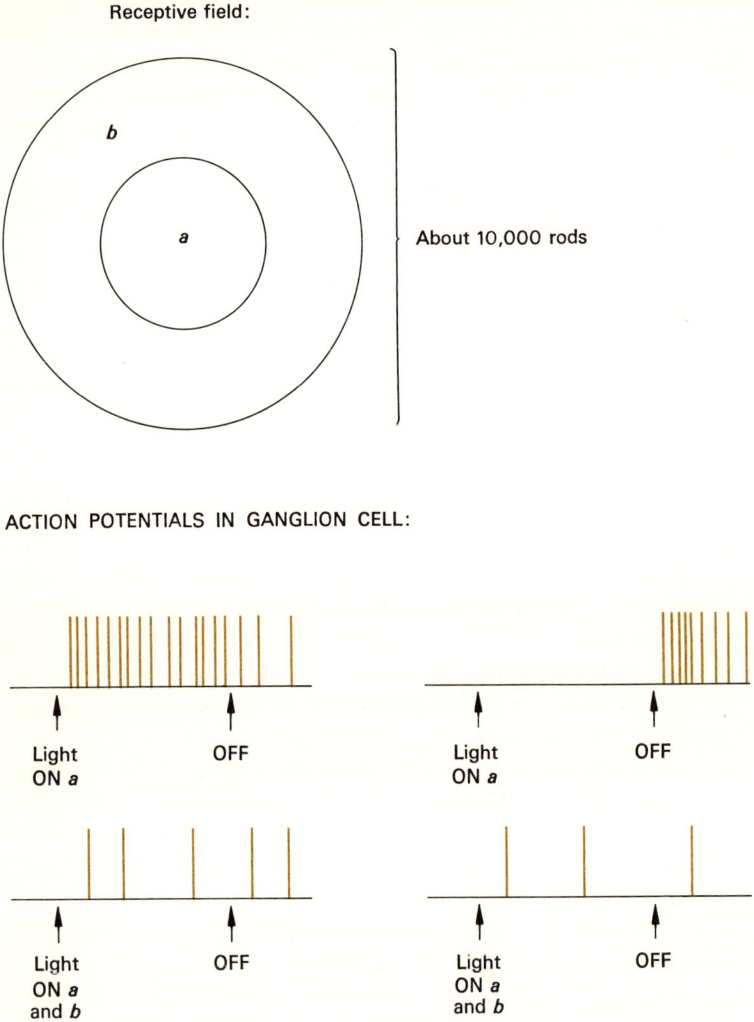

Fig. 3-26. Action potentials in a ganglion cell are evoked by illuminating receptor cells in a field of about 10,000 rods. The central part of the receptive field a *may give either an on or an off response. This is antagonized by the peripheral part* b, *as shown.*

How the region *a* grows and displaces *b* during dark adaptation (Sec. 3-4) remains a mystery, as do most of the details of signal manipulation in this "logic network."

In view of these complicated patterns it is small wonder that learning to see is a difficult process. Adults who acquire vision after a life of blindness, as through a replacement of the cornea, often become discouraged and revert to an essentially sightless way of living. Kittens who are not allowed to move about during the first few weeks of seeing and are thus prevented from coordinating vision with muscular actions behave later as if they were blind.

BIBLIOGRAPHY

1. Gregory, R. L.: "Eye and Brain," World University Library, McGraw-Hill Book Company, New York, 1966. A most entertaining account of the physiology, psychology, and esthetics of seeing.
2. Pirenne, M. H.: "Vision and the Eye," 2d ed., Chapman & Hall, Ltd., London, 1967. An excellent introduction with good historical perspective.
3. Polyak, S. L.: "The Retina," The University of Chicago Press, Chicago, 1941. A fine and thorough treatise on the anatomy of the retina.
4. *Cold Spring Harbor Symp. Quant. Biol.,* vol. 30, 1965. Papers on vision present the recent state of knowledge from a full professional point of view.
5. Giese, A. C., ed.: "Photophysiology," Vol. II, Academic Press, Inc., New York, 1964. The chapters by D. Kennedy, W. A. H. Rushton, and H. B. Barlow provide sophisticated reviews of different aspects of vision.
6. *Sci. Am.:* The frequent articles on vision are generally most interesting and informative. See, for example, E. Land, Experiments in Color Vision, May, 1959.

4.
Light as it Regulates Daily and Seasonal Rhythms Adapting to Day and Night and to the Seasons

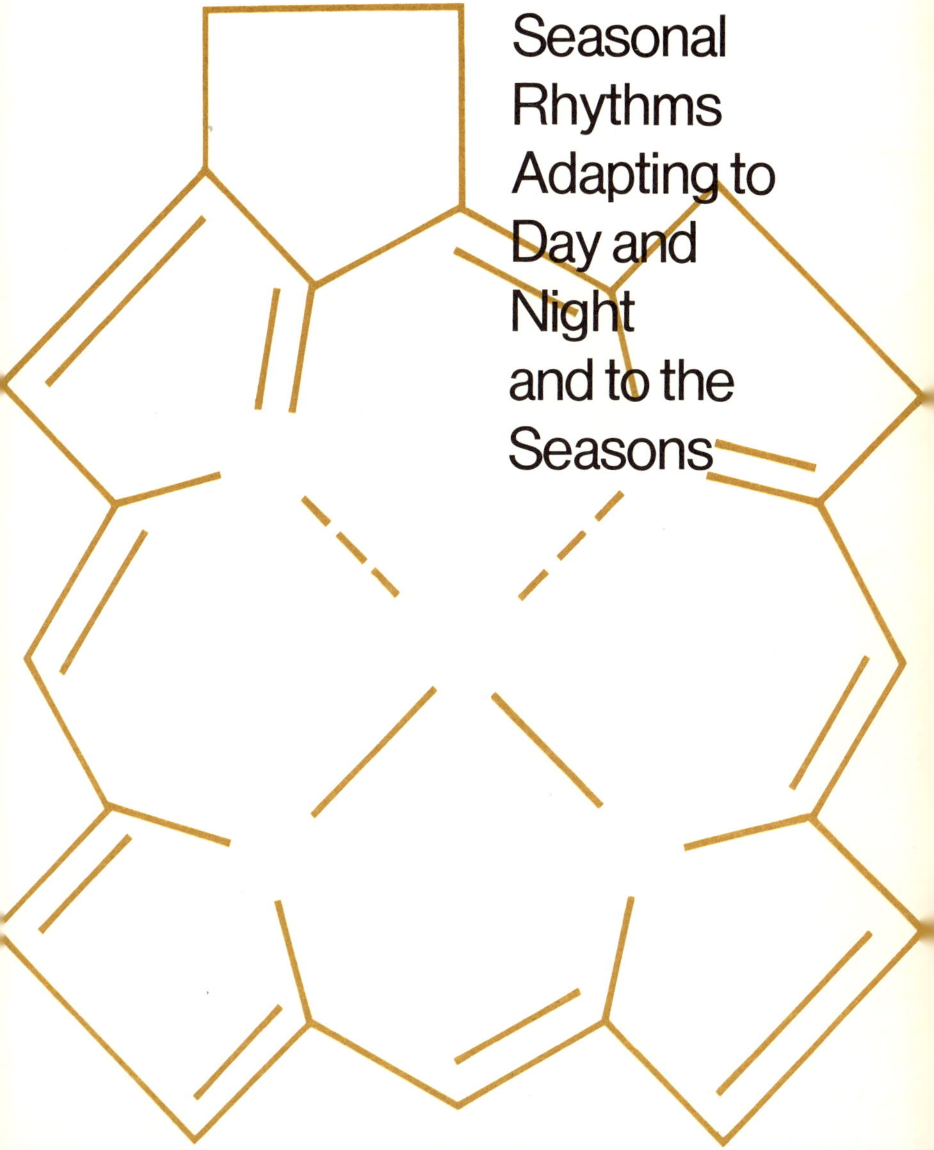

ADAPTING TO DAY AND NIGHT AND TO THE SEASONS

4-1. Daily Rhythms

Most living things are affected directly or indirectly by light, and have adapted to the cycle of day and night. A regular cycle of behavior is sometimes a direct and obvious result of the daily alternation of light and dark, as with photosynthesis or with the seeing that is needed for effective movement. A more subtle but widespread situation is that an organism has an internal clock, based on some sort of regular biochemical oscillation with a period of about 24 hours. A signal from the outside, usually the onset of dawn or dusk, keeps this clock in step with the natural day-night cycle. The clock regulates various functions of the organism.

Such endogenous clocks are called *circadian* (Latin, "about daily") to emphasize the fact that the natural period is only *approximately* 24 hours, usually between 22 and 26 hours under constant laboratory conditions. This point is important, because it shows that the clock is internal and is not a response to some aspect of the external daily cycle. When isolated from the obvious daily fluctuations of light and temperature, the clock, manifested as a rhythmic cycle of behavior, will continue day after day with its characteristic period. If this period does not happen to be exactly 24 hours, the clock will soon get out of phase with the natural (external) day.

Since light is the most spectacular, reliable, and biologically important parameter in the day-night cycle, it is not surprising that most organisms use dawn or dusk as a signal to reset their clocks each day.

The circadian clock is found in all eukaryotic[1] categories of life. Two familiar examples of its expression are (1) the daily opening and closing of flowers and leaves, and (2) the feeding

[1] Eukaryotic organisms are those whose cells have, among other things, clearly organized nuclei. This excludes the bacteria and the blue-green algae.

habits of bats, coordinated with the activities of nocturnal insects which the bats eat.

The clock can be imagined as a metabolic reaction sequence that oscillates, whose period is temperature-compensated, and whose phase can be set by a light-dark transition. The temperature compensation is only fair; the natural periods of various circadian oscillators may vary (in either direction) by about one hour for every 10°C change. This small inaccuracy is unimportant if in nature the clock is reset every day.

There are variations of behavior synchronized with more subtle changes of the environment, but these cases are rare. As examples, (1) radio waves associated with sunspots can affect the germination of seeds; (2) some algae migrate up and down in the sea (or in an enclosed tank in the laboratory) in phase with the lunar tides and rephase if moved to a new location; (3) the grunion, a species of smelt, spawns on the beaches of California at the highest tides that follow full moons in the spring, even when the moonlight is blocked continuously by clouds.

Also there are rhythms that seem not to be affected by the environment, such as the estrous cycles of many animals: about 5 days in mice, 20 in the cow, and 36 in the chimpanzee (the 28-day cycle in humans is obviously not synchronized with the phases of the moon). Again, the development of the mold *Neurospora* goes through cycles ranging from 18 to 110 hours in different laboratory strains, unaffected by light. But in other strains of the same mold the developmental cycle is circadian and responds to light.

Sometimes the direct effect of an external light-dark cycle is superimposed on a light-independent endogenous rhythm, which makes a confusing picture.

4-2. Seasonal Cycles

It is most important for animals and plants in temperate or polar regions to adapt to the seasons. Seasonal adaptation is seen in the reproductive habits of animals, in migration patterns, in cyclic molting and regrowth of hair or feathers, in cycles of activity and dormancy in both animals and plants, and in cycles of development, especially in insects and in plants.

Many of the events of seasonal adaptation take so much time that a given season must be anticipated by weeks or even months. The most reliable indicator of present and future seasons is the ever changing length of the day,[1] except in the tropics where seasons are ill defined and unimportant anyway.

By studying seasonal changes in the laboratory, under artificial light-dark cycles, we can see that most of these changes are initiated when the length of the daily light period is made greater or less than some critical value. Sometimes the control is by temperature, as in the release from dormancy of underground bulb plants, but usually the temperature has only a secondary influence.

As with the daily rhythm, the seasonal cycle involves two alternatives: First, an annual cycle might be impressed purely from the outside, by the responses to day length. Second, there may be an internal rhythm with a period of about one year, entrained to the actual seasons by a response to the day length. As an example of an internal rhythm, the growth of antlers in equatorial deer shows an annual cycle, but its phase is set by the time of birth, which can be any time in the year. This *circa-annual* rhythm is not involved with day length. But in temperate regions the cycle of antler growth is geared to the seasons through a response to the lengthening days in the spring.

Seasonal regulation based on day length is called *photoperiodism*. The triggering influence may be either one or more long days or one or more short days. Autumn flowers such as chrysanthemums are formed in response to short days (and long nights), and the development of spring flowers is triggered by long days. In aphids the short days of autumn lead to the development of sexual forms, and the long days of spring induce local migrations.

Although dominated by day length, the seasonal adaptations are usually modulated by temperature. A cold spring will retard

[1] Day length as used here will mean the duration of the light phase, in contrast to night.

the emergence of buds from dormancy or of insects from their pupae. A more unusual example bears on the apparent difficulty of inducing some African violets to flower: They flower best with a combination of cool days and warmer nights.

Latitude also modulates the response to day length. A certain butterfly living at 43° north latitude will refrain from emerging from its chrysalis and remain dormant over the winter in response to nights longer than 9 hours. For a more polar race of the same butterfly, living at 50° north latitude, the critical night length is only 6 hours. The commitment to spend the winter as a pupa is thus made a little earlier in the more northern race.

In many insects the males emerge from their pupae several weeks before the females. This ensures that only those males strong enough to survive the interval can fertilize the eggs of the females.

There are many pitfalls in the interpretation of primary and secondary causes of seasonal changes. For example, the ovaries of some birds grow in the spring, and this can be traced back to the lengthening days, but this development in the females is actually a response to the presence of sexually prepared males. The more direct effect of day length is to trigger the testicular development of the males.

In relation to these phenomena the primary photobiological questions are, How do living things measure the length of the day? What is the photoreceptor, and how does it work?

The simplest picture is that of a chemical hourglass: a reversible process that is driven in one direction by light ("setting the hourglass") and then trickles back during the night:

$$\mathrm{A} \underset{\text{dark}}{\overset{\text{light}}{\rightleftharpoons}} \mathrm{B}$$

The length of the night is measured by the extent of the back-reaction, before it is overwhelmed at dawn by the much faster light-driven forward reaction. This idea becomes less simple when we add (as we must) the feature of temperature compensation. Some migrations, as of swallows, are timed to within a few days. This means that the triggering day length has been measured with a precision of about 1 percent, in the face of wide fluctuations of the temperature.

As we shall see, the hourglass principle has been applied especially to explain the photoperiodic control of plants, less often of animals.

Another idea, advanced in the 1930s by E. Bünning and developed especially by C. S. Pittendrigh, invokes the circadian clock, already used for daily adaptation, as a tool to measure day length for seasonal control. The circadian clock is an internal cycle of chemical changes which can "tell the time." To be more graphic we can imagine the internal cycle to be divided into a "subjective day" and a "subjective night,"[1] of approximately equal lengths, partitioned by subjective dawn and dusk. Thus a colony of mice kept under constant laboratory conditions is active during the subjective day and sleeps during the subjective night. Now, under natural conditions the circadian clock is set by actual dawn or dusk. This determines the onset of subjective day or night. The day length is then appreciated by comparing the next actual dawn or dusk with its subjective counterpart. A long day can be measured by the degree to which dawn comes early, in relation to subjective dawn: the subjective night is broken prematurely. With this new dawn the clock is reset for another day.

This idea is somehow harder to translate into chemical terms than the idea of an hourglass, so there has been a tendency to resist it. But the use of *skeleton photoperiods,* employed by W. S. Hillman and others in experiments on the flowering of plants, has brought strong evidence against the hourglass concept. A skeleton photoperiod is one in which, for example, a 16-hour light period is replaced by two short flashes spaced 16 hours apart. In this case a cycle of "16-h light/8-h dark," etc., is replaced by "flash/16-h dark/flash/8-h dark," etc. When the skeleton regime and its filled-

[1] Bünning calls these the *photophil* phase and the *skotophil* phase respectively.

out counterpart have equivalent effects on flowering, it becomes difficult to cling to an hourglass model.[1] And as long as there is a circadian clock (a "pendulum" clock), why invoke an hourglass in addition? In the end we must be guided by clear specific knowledge, which remains scarce because of the complexity of the situation. Two further experiments that bear on this question will be described in Sec. 4-10.

To formulate the use of a circadian "pendulum" clock a little more precisely, we can imagine (for example) that a long-day response is triggered by a reaction that is especially sensitive to *light during the subjective night,* which results from an early dawn or a late dusk. But we must foresee the possibility of enormously complicated interactions:

1. Light sets the phase of a daily rhythm.
2. Light starts or stops a chemical hourglass.
3. Light promotes other activities such as photosynthesis and wakefulness.
4. Past light-dependent activities determine the present state and the current responses. In particular, at any instant the circadian rhythm and/or the hourglass determine the responses to light, *including* the way that light sets these timing devices.
5. These phenomena intersect with other factors such as temperature and latitude.

We can expect that these potentialities for complexity have been explored fully in the continuing experiment of evolution: "Anything that can happen will happen." Thus we have such niceties as the interactions of sunlight and moonlight to coordinate the spawning of some marine animals with the tides.

[1] Skeleton photoperiods involve subtleties that can be appreciated by considering the skeleton counterpart of an 8-h light/16-h dark period: flash/8-h dark/flash/16-h dark, etc. If this is compared with the skeleton for 16-h light/8-h dark described in the text, the two skeletons are seen to be the same when repeated a number of times, except for the starting point of the cycle. Thus, except for the starting point, the skeleton for a repeated "short day" regime can be the same as that for a repeated "long day" regime. And yet the two skeleton regimes can have opposite effects on flowering.

We should try to keep our study of photoregulation simple by examining separate aspects of the problem in systems where one or another aspect predominates.

Our present knowledge of the photochemistry of regulation in animals is so rudimentary that we shall survey only some light-dependent behavior patterns and trace some outlines of interactions between light, brain, hormones, and responses. Then we shall concentrate on the behavior of plants, in which a universally important photoregulatory pigment, *phytochrome,* is becoming well characterized.

SOME PATTERNS AND EXAMPLES OF PHOTOREGULATION

4-3. Circadian Rhythms in Animals

A partial list of daily rhythms in animals, synchronized to the natural light-dark cycle, is:

1. Feeding patterns of many animals, often interrelated. In bees the feeding is coordinated with a circadian rhythm of the opening and closing of flowers. Different flowers open at different times of the day, and the bees establish a visiting schedule accordingly.

2. Color: of shells in some crabs; of the skin in some lizards. This is achieved by the expansion and contraction of pigmented cells in or near the surface (see the *sieve effect* in Volume 1, Sec. 3-4, page 83).

3. Local migrations between sleeping and feeding areas. Pigeons can use the sun as a compass; for this they must "know" the time of day. A pigeon becomes disoriented if the phase of its clock has been shifted by an artificially early or late dawn.

4. Time of emergence of adult insects from the pupa. Insects that are active by day usually emerge in the early morning; nocturnal insects, at dusk.

5. General activity: the cycle of sleep and wakefulness. This goes with cycles of metabolic activity, levels of hormones, etc.

In man there is a daily cycle of alertness, resistance to stress, response to drugs, and other factors. Persons can generally be classified as "larks" or "owls" depending on whether their competence is greatest in the morning or in the evening. It is useful to adapt to this property, especially when the level of performance is of particular importance, as during surgery (both for the patient and for the surgeon).

The presence of two or more clocks in a single animal is sometimes suggested by changes of their relative phases. The starling shows rhythms of activity and of local migration. Both are set by the light-dark cycle. If the phase of this cycle is shifted, the clock for activity is reset in about 2 days, but the orientation (migration) takes about 4 days to rephase.

Separate clocks can become entrained to each other without the help of an external agent such as light, presumably by hormones acting as chemical messengers between the clocks.

The mechanisms of the circadian clock and its responses to light will be the subject of Secs. 4-5 and 4-8.

4-4. Seasonal Rhythms in Animals

Some conspicuous seasonal patterns in animals, regulated by day length, are:

1. Development of insects; especially the *diapause,* or suspended animation, induced by the long nights of autumn. In different insects diapause is seen in all stages of metamorphosis: egg, larva, pupa, and imago (adult). Fritillary butterflies spend the winter in the delicate condition of newly hatched larvae.

2. Migrations of insects, fish, birds, and mammals, related to reproductive cycles.

3. Patterns of reproductive competence and activity: spawning of fish, ritual combat (for breeding by the fittest) of many male animals in the mating season, development of gonads, estrus. In sheep estrus is induced by short days; in the horse, by long days.

4. Molting and regrowth of hair or feathers, with or without changes of density or color; also shedding and regrowth of antlers, and many other morphological changes.

An especially well-studied complex of photoperiodic responses in some birds is the coordinated pattern of migration, sexual activity, and metabolic variation. The lengthening days of spring lead to increased pituitary activity, gain in body weight, and fat deposition ("fueling up"), migration to a more polar region, growth of testes in the males, mating behavior and consequent growth of ovaries in the females, and nesting. Late summer or fall (short days) brings molting, migration to the winter grounds, and preparation for the responses to the next spring's lengthening days. The springtime responses cannot occur again without the intervention of the autumnal period of short days. The short days of autumn thus launch a preparation for the later responses to long days.

A curious variation is the case of transequatorial migrants, which go from a northern summer to a southern one and avoid winter. They breed in only one of the two hemispheres, and their responses to day length must obviously be different from those of more conventional animals.

The ability to respond many months after a day-length stimulus suggests the presence of an endogenous circa-annual rhythm. In a few cases such a rhythm has been confirmed, especially with birds kept continuously under an artificial 12-h daylight–12-h night cycle. Without the stimulus of lengthening days in the spring, the testicular growth response is delayed, but it eventually does occur. In nature the rhythm of this response is entrained accurately to the seasons.

4-5. Mechanisms in the Photoregulation of Animals

Action spectra for circadian clock setting and for photoperiodic responses have not shown clearly whether animals use the same or different photoreceptors for these processes. In vertebrates the receptors are in the eyes (but are not the usual visual receptors) and also in parts of the brain.

The *pineal organ* in the brain has features suggesting that it is a vestigial third eye of vertebrates. In the frog it contains cells that look like visual receptors, with multilayered membranes. Light on these cells causes the organ to generate action potentials, and the action spectrum suggests that the effective pigment is iodopsin. But in the sparrow the pineal body, although involved in a circadian rhythm of activity, does not act as a photoreceptor. Removal of the pineal body causes loss of the endogenous rhythm expressed under constant illumination, but light-dark cycles continue to impress a rhythm.

The *pituitary gland,* which regulates the output of other glands and exercises wide metabolic and sexual control in vertebrates, also plays a major part in daily and seasonal cycles. In one species of lizard the skin is green at night and brown by day. This circadian alternation continues under constant illumination unless the pituitary has been removed.

In some birds the photoreceptive regions (for both circadian and photoperiodic control) are mainly in parts of the brain near the hypothalamus. Removing the eyes or cutting the optic nerves does not reduce the responsiveness to light, but painting the top of the skull with india ink does interfere. In other birds, and in some mammals such as mice, there are receptors both in the retina of the eye (with maximum sensitivity in the red) and in the brain (maximum response to blue-violet and to red). And sometimes the photoreception is entirely in the retinas, becoming lost if the optic nerves are cut. This is the case for the ferret, in which estrus is controlled by light.

In ducks, ferrets, and probably most vertebrates, photoperiodic control involves a certain sequence of nervous and glandular events that was delineated in the Pekin duck by the careful anatomical and physiological work of J. Benoit during the 1950s and 1960s. The method was to interrupt various nerve and blood vessel connections in the brain, and to note the consequences for the testicular growth response to long days. The pattern, sketched in Fig. 4-1, is as follows: Light causes photoreceptors in the eye and in the brain to send action potentials to the *hypothalamus,* which contains secretory cells in contact with blood vessels. Incoming action potentials cause the secretory cells to send chemicals (hormones, by definition) into the blood vessels of the portal system, an intimate circulatory link between the hypothalamus and the pituitary.[1] The chemicals secreted by the hypothalamus, on reaching the pituitary, stimulate the latter organ to elaborate hormones that govern a wide range of body functions.

With this outline we can visualize in general terms how light exerts periodic control over fat metabolism, sexual activity, wakefulness, and many other aspects of metabolism and behavior.

In insects the stages of development are controlled by hormones secreted in the brain; diapause (suspension of development) results when the secretion of a particular hormone is suppressed or its transmittal from the brain is blocked. Suppression is assured by long nights; an early dawn triggers release of the hormone and further development of the insect. It is not known whether this light response is mediated by a circadian clock or by a separate hourglass. Photoreception is in the head; the most effective light is in the blue part of the spectrum.

In aphids there is no obvious circadian rhythm, and the long nights of autumn bring a change from asexual (hermaphroditic) multiplication to sexual reproduction. Here we might have an hourglass based on a sequence of events, necessary for the change to sexuality, that can be interrupted by light.

4-6. General Features of Photoregulation in Plants

In plants the measurement of day length for adjustment to the seasons involves a single ubiquitous photoreceptor pigment, phytocrome. This pigment exists in two forms: a red-absorbing form, Pr, with an absorption maximum at about 665 nm, and a far-red form, Pfr, with maximal absorption near 725 nm. The two forms are interconvertible by light, and establish a photostationary state when illuminated:[2]

$$\mathrm{Pr} \rightleftharpoons \mathrm{Pfr} \tag{4-1}$$

[1] The pituitary is also called the *hypophysis.* It is made up of a lobe of nerve tissue (the neurohypophysis) and one of glandular tissue (the adenohypophysis).

[2] Recall the dynamics of a photostationary state, described for rhodopsin in Sec. 3-6 (see Fig. 3-20).

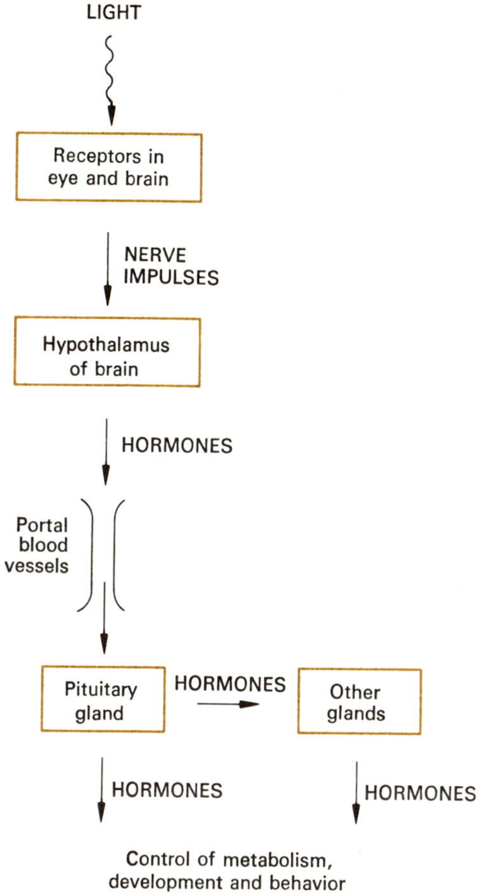

Fig. 4-1. A sequence of events in the photoperiodic regulation of animals, as exemplified by the action of long days on sexual maturation of Pekin ducks.

Red light, absorbed mainly by Pr, tends to shift the equilibrium toward Pfr, and vice versa for far-red light. Thus red and far red have opposite effects on the relative concentrations of Pr and Pfr.

Historically the photoperiodic function of phytochrome was

implicated first by the fact that seasonal adjustments are regulated by light, second by action spectra showing the importance of red and far red, and third by the frequent observation that far-red light undoes the effect of red light and vice versa. Finally the pigment itself was isolated and could be studied in pure form, as we shall see in the next section.

The importance of day length in triggering seasonal changes emerged in the 1910s and became well documented by W. W. Garner and H. A. Allard, who showed that the length of day (or of night) was critical in governing the development of flowers in many plants. Because of a general prejudice that flowering should be controlled by temperature, Garner and Allard had difficulty at first in having their reports accepted for publication.

Garner and Allard showed that there are *short-day* and *long-day* plants, flowering in autumn and spring respectively, and *day-neutral* plants, in which flowering is independent of day length. Their work was extended by H. A. Borthwick, S. B. Hendricks, and M. W. Parker, who showed that other responses such as seed germination are regulated by light and who consolidated the evidence for antagonistic effects of red and far-red light.

A typical experiment concerns the flowering of soybean plants, triggered by a single night longer than about 14 hours. This flowering is prevented if the long night is interrupted somewhere near its middle by a flash of light. A single flash of red light of about 10^{-2} watt/cm^2 lasting 1 sec (total, 10^{-2} joule/cm^2, or 10^5 erg/cm^2) is enough to prevent flowering. The effect of the red flash is erased by a far-red flash of somewhat greater strength. A sequence of alternating red and far-red flashes prevents flowering only if it ends with a red flash. Also a far-red flash can substitute, in part, for a longer night in triggering the flowering response. Action spectra for the effectiveness of the red flash that inhibits flowering and of the far-red flash that erases the red flash show maxima near 665 and 725 nm respectively.

The fact that far red can substitute in part for a longer night suggests that far red and darkness do the same kind of thing: cause or allow the conversion of Pfr to Pr:

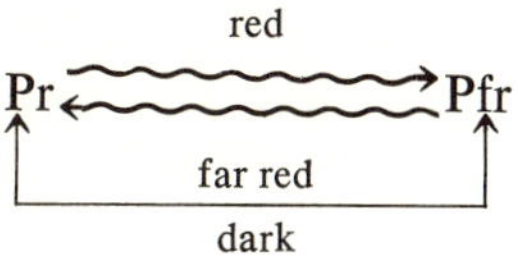

(4-2)

This is a model for an hourglass. During the day a photostationary state is maintained; the equilibrium favors Pfr. At dusk the hourglass is set: photoconversion ceases, and the slower "dark" trickle of phytochrome from the Pfr form to Pr begins. If the night is long enough, the concentration of Pfr falls below some critical level. A flash of red light regenerates some Pfr and "sets back" the hourglass. Conversely a flash of far red can substitute for a long night by quickly converting Pfr to Pr. These interpretations are outlined in Fig. 4-2.

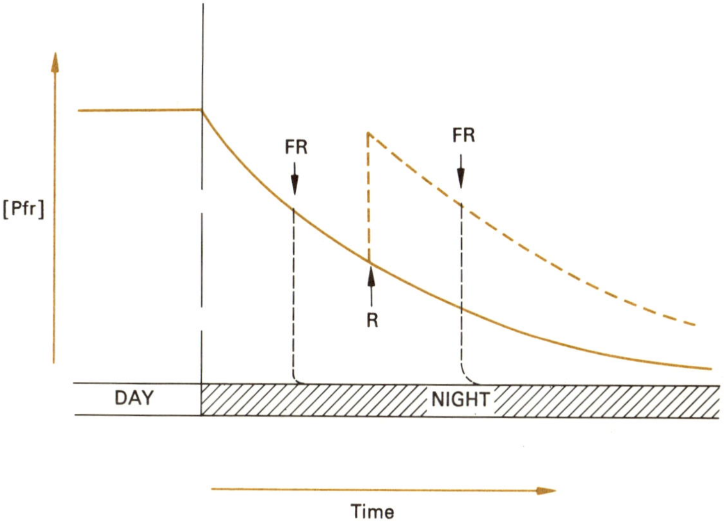

Fig. 4-2. The predicted level of the far-red absorbing form of phytochrome, Pfr, is plotted against time. The balance between Pr and Pfr is dictated during the day by a photostationary state (Pr⇌Pfr). At night the reaction Pfr → Pr could act as an hourglass (solid curve). A red flash (R; dashed curve) retards the hourglass by converting some Pr back to Pfr. A far-red flash (FR; dotted curves) advances the hourglass by a rapid conversion of Pfr to Pr.

The antagonistic effects of red and far-red light are seen in many aspects of plant development and behavior. In some cases the day length does not need to be involved; for example, the germination of many seeds simply requires light. A weak flash of red light will do, and is erased by a comparable flash of far red. Again, the circadian rhythm of leaf movements can be affected by the antagonistic effects of red and far-red light. The "sensitive plant," *Mimosa pudica,* folds its leaves in response to a transition from light to darkness. This response is inhibited if a flash of far red is interposed between the end of the light period and the start of the dark period. The inhibition is erased by a subsequent red flash, restored by another far-red flash, and so forth. These effects are interesting because the levels of Pr and Pfr cause responses that can be seen within 5 min.

Other rapid effects apparently mediated by phytochrome are (1) the red-accelerated and far-red-retarded growth of seedlings, (2) some effects of red and far red on phototropic sensitivity of seedlings, and (3) effects of red and far red on the phototaxis of chloroplasts which align themselves edgewise to strong light and facewise to weak light (see Chap. 2).

Other effects mediated by phytochrome do involve the day length. A list of these photoperiodic responses includes:

1. Onset of dormancy (short-day response) and release from dormancy (long-day) in buds, evergreens, and some bulbs (but in many bulbs that are entirely underground in the winter, the release is induced by temperature and not light)
2. Flowering, both short-day and long-day
3. Synthesis of anthocyanin pigments and degradation of chlorophyll in the leaves of deciduous trees (short-day), giving the autumn colors

To establish a photostationary state between Pr and Pfr it should be enough if every molecule of phytochrome absorbs several quanta. For most pigments, including phytochrome, about 10^{16} quanta/cm^2 (see Probs. 9 and 10, Volume 1) are sufficient. With red light this amounts to about 3×10^4 ergs/cm^2. This explains why such weak flashes of red and far-red light are effective in interrupting a long night. There are other effects of light on plant development that require many orders of magnitude more energy, and are therefore lumped together under the name *high-energy reactions*. Because the total energy delivered in one day depends on the day length, these high-energy reactions can influence the seasonal patterns of development. High-energy reactions are seen in the light requirements for maximal stem and leaf growth (aside from the dependence on photosynthesis), for anthocyanin synthesis, and for the formation of tubers in some plants such as *Begonia*.

The high-energy reactions generally are promoted by both blue and far-red light, whereas red has little effect. These responses might involve the photodestruction of Pfr, sensitized directly by Pfr or indirectly by pigments that absorb in the blue. Photodestructive reactions will be described in Chap. 5.

We do not know the mechanism(s) by which phytochrome acts. Current speculations on this subject have been inspired by similar speculations about vision: A photochemical reaction of phytochrome leads to a change in the property of a membrane; this changes the distribution of hormones; and so forth. We shall return to these matters in Sec. 4-8.

The investigators who have been preoccupied with phytochrome and photoperiodism in plants have tended to ignore the existence of circadian rhythms, and to discount the possibility that an hourglass based on Pfr $\longrightarrow$ Pr might be replaced or modified by some other internal clock.

The observation of rhythms in plants goes back to Androsthenes, who recorded daily leaf movements (the "sleep of plants," according to the natives of the island Tylos) while marching with Alexander the Great. The persistence of the rhythmic opening and closing of leaves and flowers under constant illumination has been known for more than two centuries, and the ability of light to change the phase of the rhythm was noted by de Candolle in 1832. We now recognize many light-regulated circadian rhythms in plants, algae, and protozoa:

1. Opening and closing of leaves, flowers, and stomata (the pores of leaves that control gas diffusion)
2. In one-celled algae and photosynthetic protozoa, the ca-

pacities for photosynthesis, phototaxis, and luminescence; also cell division and vertical migration in the sea

3. The ability to respond to the photoperiodic stimuli mediated by phytochrome, such as the interruption of a long night by a flash

The circadian and the seasonal photoperiodic systems interact in ways that can become quite complicated and confusing. Only recently have a few scientists attacked this difficult aspect of photoregulation.

Some relationships between circadian and seasonal rhythms, or between endogenous clocks and the physiology of phytochrome, will be examined in Sec. 4-10. First we must develop our picture of the photochemistry and physiology of phytochrome on one hand and of the circadian clock on the other.

PHYTOCHROME

4-7. The Chemistry of Phytochrome

The opposite effects of red and far-red light in the control of plant development led scientists in the 1950s to search for a pigment (phytochrome) that could be interconverted by light between red-absorbing and far-red-absorbing forms. Conversion from Pr to Pfr by red light should be manifested by an increase in the optical density at 725 nm and a decrease at 665 nm. Far-red illumination should cause changes in the opposite direction.

Certain difficulties could be anticipated. The concentration of the pigment might be very low, giving a minute change of OD in a strongly scattering piece of plant tissue. Absorption and fluorescence by chlorophyll could interfere seriously with the intended observation, so a "white" tissue should be preferred. Young dark-grown seedlings could be tried, but in these the light-stimulated formation of chlorophyll from protochlorophyll could give an increased OD in the red that might be confused with the formation of Pr. Eventually parsnip roots and cauliflower heads proved to be highly suitable materials for the detection of phytochrome in vivo, and oat seedlings were found to be a good source of the pigment for its isolation and purification.

Light-induced changes of OD revealing the conversion (and thus the presence) of phytochrome in living tissues were first reported in 1959 by W. L. Butler and associates. They found changes in OD of about 0.01 in seedlings and parsnip roots exposed to cycles of red and far-red illumination. These changes could be used as an assay for phytochrome, to monitor the success of attempts at purification. As a result H. W. Siegelman and others were able to extract phytochrome from seedlings, purify it, and show that it is a protein of molecular weight 60,000 [1] with one or more chromophores giving the pigment a blue color as Pr and green as Pfr. The purified phytochrome could be converted repeatedly from Pr to Pfr and back. Pfr also returns to the form Pr in the dark, in keeping with the hourglass model for its action in the plant, but the return in vitro is quite slow: return of half of the material takes about 10 hours at room temperature.

In seedlings about 0.1 percent of the total protein is phytochrome.

Optical properties of phytochrome are shown in Fig. 4-3. The top curves show action spectra for the induction (by red light) and inhibition (far red) of lettuce seed germination. The middle curves show action spectra for the conversion of Pr to Pfr and vice versa, measured with purified phytochrome. The bottom curves show absorption spectra of phytochrome after far-red and red illumination respectively. After far-red illumination the absorption spectrum reflects almost pure Pr, but after red illumination (660 nm) the photostationary state is about 80 percent Pfr and 20 percent Pr. We shall return to this point. There are lesser absorption maxima in the violet, at 375 nm for Pr and 400 for Pfr.

The action spectra indicate that the response in lettuce seeds is indeed mediated by phytochrome.

The absorption spectrum of Pr, when compared with that of phycocyanin, suggests that the chromophore of phytochrome is similar to that of phycocyanin: an open-chain tetrapyrrole, some-

[1] For the pigment derived from oat seedlings.

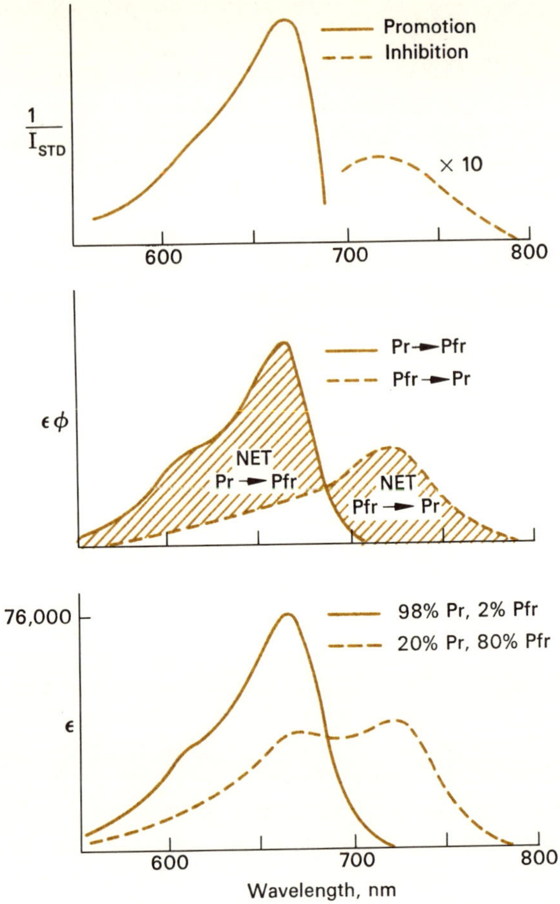

Fig. 4-3. Optical properties of phytochrome. Top curves, action spectra for promotion and inhibition of the germination of lettuce seeds (reciprocal of the quantum intensity needed for a standard response). Middle curves, effectiveness spectra for interconversion of the two forms of purified phytochrome, Pr and Pfr. The effectiveness is the product of the extinction coefficient of either form and the quantum efficiency for conversion to the other form. Bottom curves, absorption spectra of phytochrome. Solid curve, about 98 percent Pr, produced by irradiating a solution of phytochrome with 730 nm light. Dashed curve, 80 percent Pfr and 20 percent Pr, produced by irradiation with 660 nm light.

what like a chlorophyll molecule that has been snipped and stretched out in a line. It has been difficult to prepare even minute amounts of the chromophore separated from the protein; a kilogram of seedlings gives a few micrograms of chromophore. But from partial information about spectra and other physical and chemical properties H. W. Siegelman has proposed the chemical structure shown in Fig. 4-4.

The photostationary state attained by a pigment such as phytochrome during illumination depends on the rate at which each pigment absorbs light and on the quantum efficiency for the conversion of each form into the other. If these quantum efficiencies

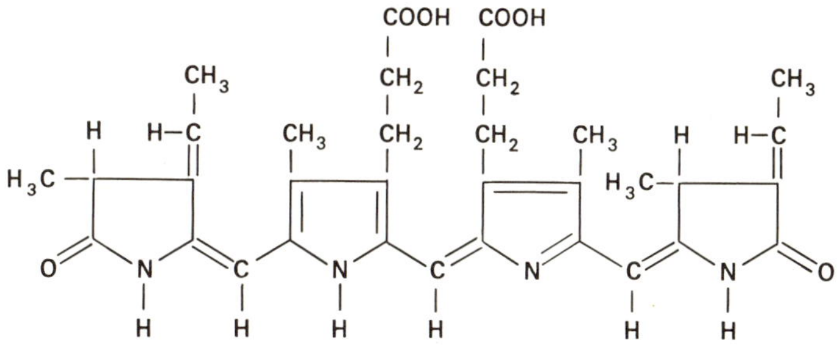

Fig. 4-4. A provisional structure for the chromophore of phytochrome. This bilitriene structure was proposed by H. W. Siegelman on the basis of similarity to the chromophore of allophycocyanin.

are ϕ_r and ϕ_{fr} (for Pr $\longrightarrow$ Pfr and Pfr $\longrightarrow$ Pr respectively) and if ε_r and ε_{fr} are the extinction coefficients of Pr and Pfr at the wavelength of illumination, the ratio of Pfr to Pr is

$$\frac{[\mathrm{Pfr}]}{[\mathrm{Pr}]} = \frac{\phi_r\,\varepsilon_r}{\phi_{fr}\,\varepsilon_{fr}} \tag{4-3}$$

At 660 nm the phytochrome is about 80 percent Pfr, and under 730 nm light it is about 98 percent Pr and only about 2 percent Pfr.[1] At their respective absorption maxima the extinction coefficients of the two forms are $\varepsilon_r = 76{,}000M^{-1}\ \mathrm{cm}^{-1}$ at 665 nm, and $\varepsilon_{fr} = 46{,}000M^{-1}\ \mathrm{cm}^{-1}$ at 725 nm. The ratio of quantum efficien-

[1] This ratio is given by the middle curves in Fig. 4-3. It is equal to the relative heights of the solid and the dashed curve at any wavelength.

cies, ϕ_r/ϕ_{fr}, is 1.5. These data are important, because they allow us to predict the relative concentrations of Pr and Pfr in living plants under illumination of different wavelengths.

The change from Pr to Pfr and back entails modifications of both the chromophore and the protein. These changes are still matters of speculation. Three possibilities for the change of the chromophore are shown in Fig. 4-5: internal H-shift, oxidation-reduction,[1] and isomerization. Any of these could account for the difference in the absorption spectra of Pr and Pfr, through changes in the pattern of conjugated double bonds. The first of these three possibilities is currently the most popular; the third would make the closest analogy to the photochemistry of vision.

Changes in the protein part of the molecule are shown by the fact that the protein behaves differently in Pr, in Pfr, and in intermediate states during interconversion. Thus urea denatures the protein of Pfr but not of Pr. Trypsin, a protein-digesting enzyme, attacks neither Pr nor Pfr, but it does attack the phytochrome during its conversion from Pr to Pfr. It does not attack during the conversion from Pfr to Pr, showing that the two directions of interconversion follow different paths.

These observations tend to identify Pfr and its precursors in the change from Pr as the most labile, or reactive, forms.

The interconversion between Pr and Pfr is strongly reminiscent of the chemical cycle of rhodopsin, especially in the squid, where rhodopsin and metarhodopsin A are interconverted by light and can coexist in a photostationary state. If Pr or Pfr is cooled to liquid nitrogen temperature, illuminated, and then warmed, a sequence of events is revealed that is much like the sequence of chemical steps in vision. An outline of the sequence as reported by L. H. Pratt and W. L. Butler is shown in Fig. 4-6. This picture is conservative in that the cycle probably involves more steps than the ones shown.

By studying rapid changes in absorption spectra of phyto-

[1] The oxidation-reduction process need not involve an external substrate; H atoms could be exchanged with a neighboring chromophore or part of the protein.

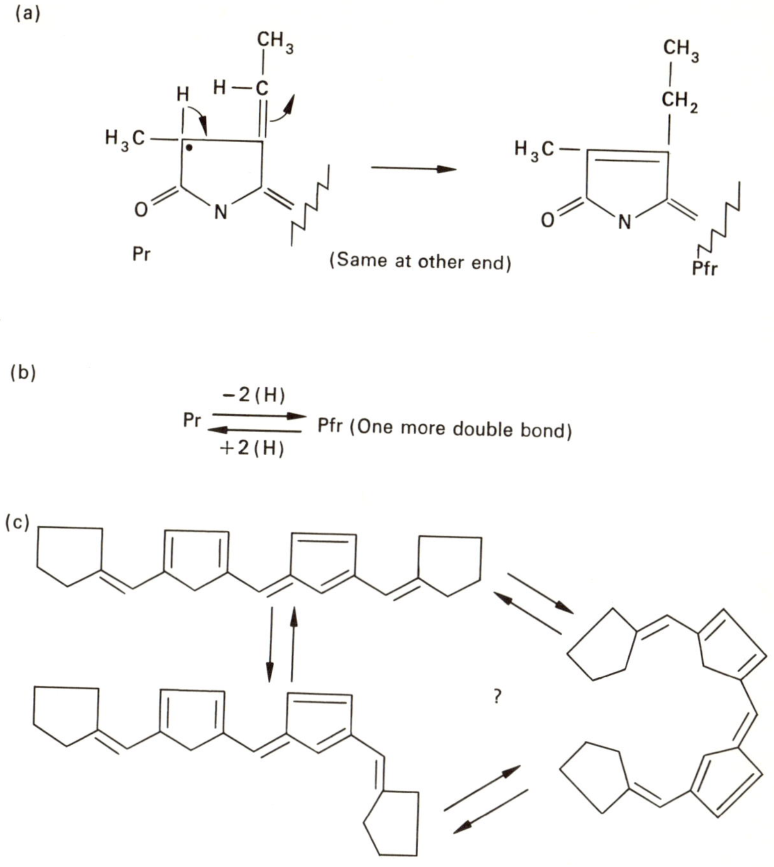

Fig. 4-5. Three hypotheses for the difference between Pr and Pfr: hydrogen shift, oxidoreduction, and isomerization.

chrome and rhodopsin after short flashes of light, with time discrimination approaching microseconds, H. Linschitz and others have gained evidence of multiple parallel pathways: perhaps three in the steps from R_{695} to Pfr, two from Pfr to FR′, and three between prelumirhodopsin and free retinal in the case of the visual pigment.

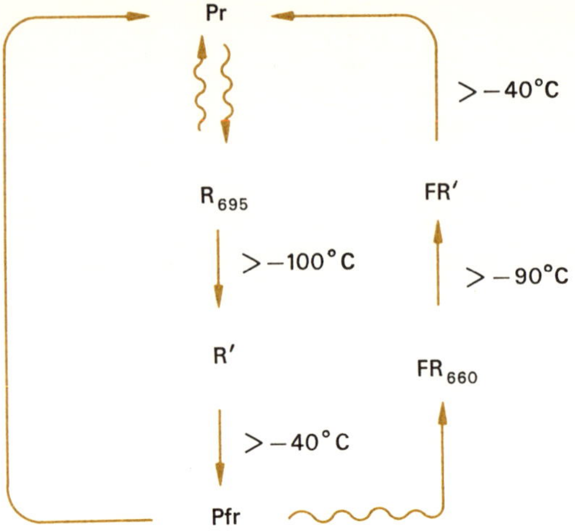

Fig. 4-6. A partial description of steps in the interconversion of Pr and Pfr, showing similarity to the chemistry of rhodopsin. Light-driven steps are indicated by wavy arrows.

4-8. The Photophysiology of Phytochrome

As mentioned before, the similarity between phytochrome and rhodopsin has encouraged certain speculations about the mechanism of action of phytochrome, as follows: The pigment-protein is associated with a plant membrane. Conversion from Pr to Pfr or vice versa exposes or covers a pore or the active site of an enzyme. This leads directly to electroosmotic changes due to currents of ions and water.

The best evidence for this picture is the effect of light absorbed by phytochrome on leaf closing in *Mimosa.* The sensitivity to red and far-red light is greatest, in this case, in an organ at the base of each leaflet. This organ is known to change its osmotic properties, and hence the turgor of the stem joint, so as to cause folding of the leaf. Effects of phytochrome-absorbed light on phototropic and phototactic responses in plants might also be traced directly to chemiosmotic changes in the membranes involved in the responses.

We can go on to imagine that the state of a membrane, affected by phytochrome, governs the distribution of various plant hormones: the auxins, which regulate growth; the kinetins, gibberellins, and dormins, which influence dormancy, germination, cell division, and differentiation. These hormones may act by determining what genes are active and thereby what enzymes are being made. Thus the state and the reactions of phytochrome can have the most pervasive effects on the life of a plant.

In seeking to understand this complicated network, investigators have looked for correlations: between the light program, the levels of Pfr and Pr, the levels of identifiable hormones and enzymes, and the final responses such as flowering. This effort has been sometimes rewarding but usually frustrating. Simple and general correlations are scarce, but there is (unfortunately) a universal and untestable excuse: If a response fails to bear an expected relationship to the concentration of Pfr or Pr, it is because a small component of "active" phytochrome has different surroundings and hence different properties from a larger component of inactive phytochrome. The observed concentrations of inactive Pr and Pfr might mask quite different concentrations of the active fraction. Similarly a failure to correlate levels of Pr and Pfr with levels of hormones or to correlate hormones with responses can be excused on the valid grounds that we have not distinguished between a number of closely similar hormones with different actions.

One simple and general idea stemmed from the apparent importance of night length as contrasted with day length. In the control of flowering a long night could be nullified by a single short interrupting flash, but a long day could not be interrupted effectively by a short dark interval. Interpretation of these facts, in terms of probable levels of Pr and Pfr, led to the thought that the active principle is Pfr. The effect of the long night, whether it be the promotion or the continued suppression of an activity such as flowering, is achieved when Pfr reaches a certain critically low concentration (see Fig. 4-2).

The idea that Pfr in small amounts is influential was reinforced by observations of the effects of red and far-red light on the growth of pea leaves. Red light, which converts some Pr to Pfr, stimulates growth. Starting with (presumably) 100 percent Pr, one can apply a flash calculated to convert 0.1 percent of the pigment to Pfr. This should produce a large *relative* change in Pfr, from zero to 0.1 percent, but only a miniscule relative change (from 100 to 99.9 percent) in the concentration of Pr. Such a flash causes a conspicuous increase in the rate of growth.

The simple picture of an hourglass based on Pfr $\longrightarrow$ Pr, with Pfr the active form, could be tested in a variety of situations. Often the picture accounted for the observations, and often it did not. For the flowering of many short-day plants this model seemed sufficient as long as the question of circadian rhythms was avoided (see Sec. 4-10).

In many plant tissues the formalism of light and dark conversion represented by Reaction (4-2), on which the hourglass model has usually been based, is incorrect. We now know, largely from studies by W. R. Briggs, that Pfr often disappears without a corresponding rise of Pr. Apparently the destruction of Pfr, or its conversion to an inactive form,[1] vies with the dark return to Pr. A more complete scheme would be

$$\text{Light:} \qquad \text{Pr} \rightleftharpoons \text{Pfr}$$

$$\text{Dark:} \quad \begin{cases} \text{Pfr} \longrightarrow \text{X} \\ \text{Pfr} \longrightarrow \text{Pr} \end{cases} \quad \text{and} \quad \text{Y} \longrightarrow \text{Pr} \qquad (4\text{-}4)$$

[1] The operational test for Pfr is to illuminate a plant tissue with far red and look for a decrease in the OD at 725 nm. If the Pfr should lose its ability to be photoconverted to something else, it would seem to be no longer there in the terms of this assay.

The reaction Y ⟶ Pr represents the synthesis of new Pr to compensate for the loss of Pfr. In peas this new synthesis is triggered when the concentration of phytochrome falls below a certain level.

Plants vary in the relative amounts of destruction of Pfr and reversion to Pr. In leaves of oats and corn the decline of Pfr in the dark is due almost entirely to destruction. In peas and many grasses the fate of Pfr is about 80 percent destruction and 20 percent reversion, and in cauliflower heads there is very little destruction.

The destruction of Pfr requires conditions that allow respiration, and it follows zero-order kinetics (the rate is independent of Pfr concentration). In contrast the reversion to Pr is independent of factors such as oxygen that affect respiration, and it follows first-order kinetics (rate proportional to Pfr concentration). Knowing this one can more easily distinguish the two processes and influence them differently.

Even allowing for these variations on the earlier and simpler picture, it is often impossible to explain photoperiodic responses simply in terms of Pfr concentration. For example, a certain predicted level of Pfr and of Pr can be established in either of two ways: one can give a flash of red or far-red light after some particular history of light or darkness, or one can set up a predictable photostationary state. These two ways of reaching a single (presumed) state of Pfr and Pr sometimes have the same consequences for flowering or seed germination, but often they do not.

In conclusion the hourglass based on a decline of Pfr in the dark remains an attractive idea, but it can often be neither proved nor disproved, and we should consider the possible involvement of a circadian clock in the measurement of day (or night) length.

RHYTHMS IN PLANTS

4-9. The Nature of the Circadian Clock

Many features of circadian clocks are seen in the behavior of the marine dinoflagellate *Gonyaulax,* as described by J. W. Hastings and B. M. Sweeney. This microscopic, armor-plated, photosynthetic, luminescent protozoan exhibits several circadian rhythms, entrained to the day-night cycle. Three are shown in Fig. 4-7: capacity for photosynthesis, maximal in the day; capacity for flashing luminescence (induced by shaking), maximal at night; and cell division, which occurs just before dawn. A weak, steady luminescent glow also is maximal just before dawn.

If a laboratory culture of *Gonyaulax* is maintained in constant dim light, these rhythms continue, with a period between 22 and

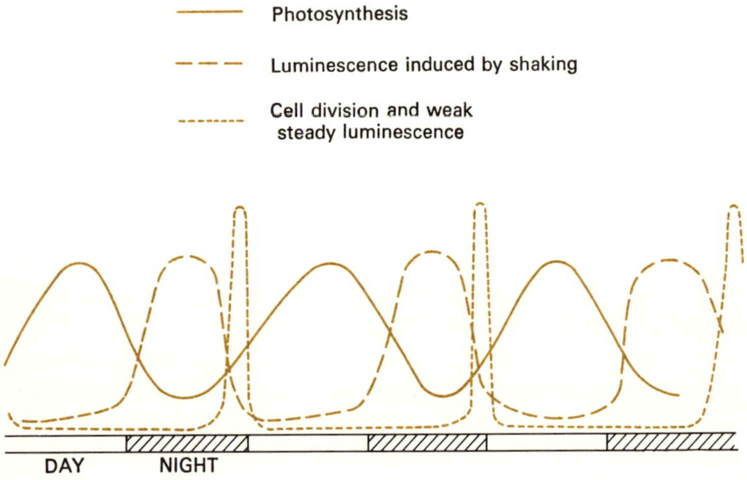

Fig. 4-7. Circadian rhythms exhibited by Gonyaulax, *a marine protozoan.*

25 hours depending on the temperature. If there is any variation among individuals, this is not apparent, but the individual cells could entrain each other in the culture. A colony of mice in a constant environment shows, as a whole, a circadian rhythm of activity and sleep with a single definite period. If single mice are isolated, they show variations up to about 20 min in their circadian periods. This would be enough to desynchronize the colony in a few days, but when together the mice probably wake each other up and keep in step.

Returning to the culture of *Gonyaulax,* it loses the expression of rhythmic behavior in bright light, but this does not necessarily mean that the underlying clock has been lost. If a culture is kept in bright light for 3 years, through about 1,000 generations, during which it shows no rhythmicity, the rhythms reappear as soon as the culture is returned to dim light. The phase of each rhythm is related to the time of the bright-to-dim transition as it would be related to dusk under a natural day-night cycle.

The rhythms of *Gonyaulax,* as in circadian systems generally, can be entrained to light-dark cycles with periods ranging from about 18 to 30 hours. If one tries to synchronize the system to shorter or longer cycles, the entrainment will break down, and the rhythms may then exhibit multiples or submultiples of the impressed period.

All efforts to change the natural period of a circadian clock, by prolonged entrainment to a period other than the natural one, have failed. After many generations on an unnatural light-dark cycle (such as 20 hours), a culture of *Gonyaulax* will revert immediately to its period of 22 to 25 hours when the entraining influence has been lifted. Nevertheless the longer evolutionary process in nature might have selected, from a wide range of rhythmic periods, those that are circadian.

Figure 4-8 shows how circadian clocks are reset by light during the subjective night. An early dawn advances the phase, and a late dusk retards it. Early dusk and late dawn advance and retard the phase similarly (not shown). The curves as drawn could represent the capacity for photosynthesis in *Gonyaulax* or for phototaxis in the alga *Euglena* (an object of extensive study by V. G. Bruce). Phase shifting can be induced by short flashes as well as by changes in the pattern of continuous light. This opens the door to the measurement of action spectra for phase shifting.

The action spectrum for phase shifting in *Gonyaulax* shows sharp peaks at 475 and 650 nm, in rough (not very close) agreement with the absorption due to the mixture of photosynthetic pigments, Chl *a* and *c* and carotenoids. In some higher plants the action spectra for phase shifting are roughly "chlorophyll-like."

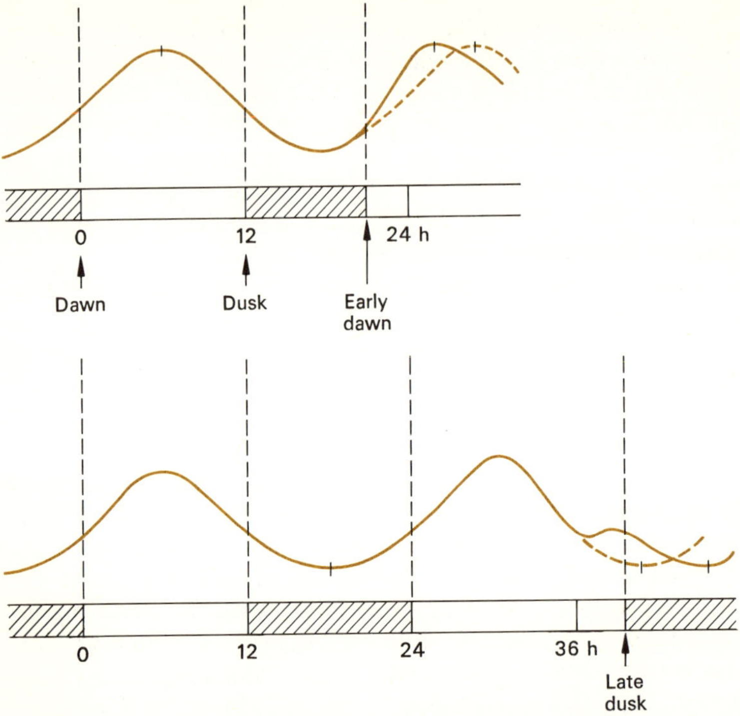

Fig. 4-8. Two ways that the phase of a circadian rhythm is shifted by light.

Sometimes, as with the rhythm for leaf movement in bean plants, light absorbed by phytochrome appears to be involved. However, this may be a case of confusion where phytochrome affects the mechanism for leaf movement rather than the phase of the underlying clock.

In a collection of fruit fly pupae, a single flash of light induces some to emerge as adults. Under subsequent darkness a second wave of emergence comes about 24 hours later, a third after about 48 hours, and so forth. The effectiveness of this false dawn is greatest with blue light; red is ineffective. A blue-but-not-red action spectrum is also found for circadian phase shifting in many insects and in the mold *Neurospora.* This enlarges our collection of "blue light effects," which includes phototaxis of green algae, phototropism of fungi and higher plants, and many other photobiological responses. As usual we resort to the cliché that the photoreceptive pigments might be carotenoids or flavins.

Many experiments implicate DNA and RNA in the working of the circadian clock, and this has led C. H. Ehret and others to propose models based on the functions of these substances. One

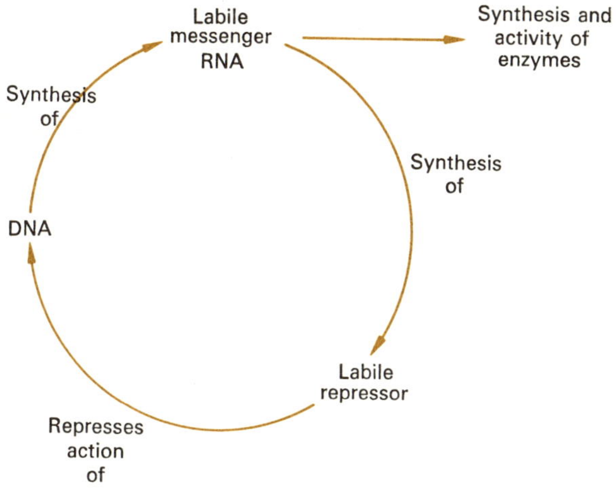

Fig. 4-9. A model for a biochemical clock (a cycle capable of oscillating) involving DNA-directed synthesis of RNA.

such model, suggested by J. W. Hastings, is outlined in Fig. 4-9. A detailed kinetic analysis shows that it should be capable of giving temperature-compensated oscillations in the rates of enzyme syntheses. Four observations that support models of this kind are:

1. A low concentration of actinomycin D ($2 \times 10^{-8}M$), which inhibits the DNA-dependent synthesis of RNA, when added to *Gonyaulax* allows one more cycle and then stops the clock. It does not inhibit photosynthesis or luminescence, only the rhythmic variations thereof.

2. Ultraviolet irradiation retards the phases of rhythms in *Gonyaulax* and in *Paramecium,* and subsequent white light undoes the effect of the ultraviolet. This looks like a case of photoreactivation (see Chap. 5), which usually involves damage and repair of nucleic acid bases.

3. A nerve in the sea hare, a large flabby marine animal, shows a burst of action potentials at dawn. The injection of actinomycin D at night gives an immediate burst of action potentials and another one about 24 hours later (the animal being kept in darkness). It could be argued that the actinomycin D strips all newly formed RNA from its DNA template and thus discharges one step in a biochemical clock.

4. The importance of the cell nucleus is shown by microsurgical experiments made with the giant algal cell *Acetabularia.* This alga shows a circadian rhythm of capacity for photosynthesis. If two cells are entrained by different light-dark cycles so as to have opposite phases, interchanging their nuclei causes their phases to be exchanged. The nucleus thus carries information about the phase of the clock.

4-10. The Clock and the Estimation of Day Length by Plants

There are many complicated ways that light, Chl, phytochrome, and circadian clocks can interact in plants. We shall sidestep the confusion and examine two simple experiments that show dramatically how circadian rhythms affect photoperiodic control.

Figure 4-10 shows an experiment on the flowering of duckweed, a long-day plant studied by W. S. Hillman. Normally the plants are prevented from flowering by long nights and can be induced to flower if a weak flash of red light interrupts the long night near its middle. A far-red flash erases the effect of the red flash. In the experiment of Fig. 4-10, plants were exposed to a long period of darkness, up to several weeks, after a light phase. For different plants the dark was interrupted, at various times after its initiation, by a red flash. Later the plants were placed on a long-night–short-day cycle and their subsequent flowering was recorded.

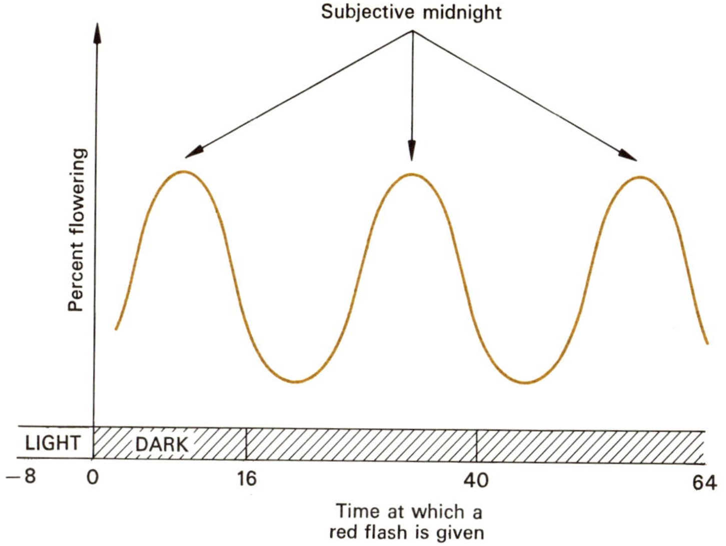

Fig. 4-10. Duckweed plants, normally long-day, can be induced to flower by a flash of red light that interrupts a long dark period. The responsiveness shows a circadian rhythm.

The figure shows that the inducing effect of the red flash followed a regular rhythm, with about a 24-hour period. The effect was greatest near subjective midnight. The rhythm could be traced through weeks of darkness, showing that a red flash can spoil a very long dark period if it is applied at the right time in relation to a circadian rhythm. This experiment shows that a simple hourglass model is inadequate to explain the control of flowering. The level of Pfr, declining in the dark and rising after a red flash, may be important, but the consequences of a certain level of Pfr are dictated by other factors controlled by a clock.

A different kind of experiment, which can be made with some short-day plants such as soybean, is shown in Fig. 4-11. A long night triggers flowering, but only if the night ends near the subjective dawn. After an 8-hour day, a night lasting 16, 40, or 64 hours induces flowering, but one that lasts 28 or 52 hours does not. Thus a very long dark period is counted as a short night if it ends during the subjective dusk or evening of a circadian cycle.

One can make a variety of hypotheses to explain these and other more complicated observations. A simple assumption could be that phytochrome sets the activity of a certain enzyme but a circadian rhythm determines the availability of the substrate of that enzyme (or vice versa). As long as light can influence the

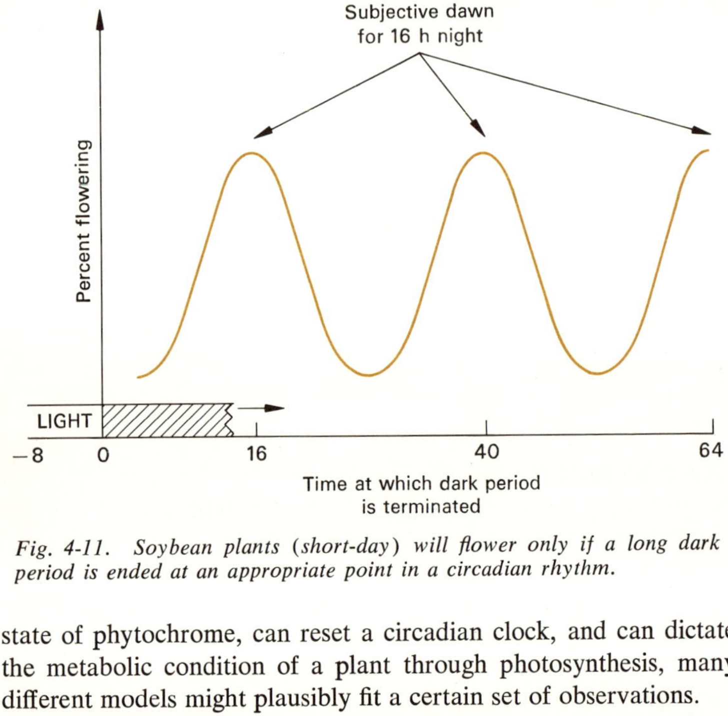

Fig. 4-11. Soybean plants (short-day) will flower only if a long dark period is ended at an appropriate point in a circadian rhythm.

state of phytochrome, can reset a circadian clock, and can dictate the metabolic condition of a plant through photosynthesis, many different models might plausibly fit a certain set of observations.

The variety of regulatory devices found in nature is marvelous, and shows the need for continued investigation with careful attention to the problem of separating the variables. Our picture of the physiology of photoregulation has passed from one of relative simplicity into that confused state that often precedes a highly satisfying clarification at a new level of understanding. This should be a rewarding area of research.†

† Since this was written, evidence has emerged that the protein of phytochrome is changed drastically during its extraction and purification.

BIBLIOGRAPHY

1. Hendricks, S. B.: How Light Interacts with Living Matter, *Sci. Am.,* vol. 219, September, 1968, pp. 175–186. A short discussion of phytochrome and photoperiodism in plants, along with other topics in photobiology.
2. Reinberg, A., and J. Ghata: "Biological Rhythms," English translation from the French, Walker and Co., New York, 1964. A simple but sound description of the subject.
3. Sweeney, B. M.: "Rhythmic Phenomena in Plants," Academic Press, Inc., New York, 1969. A short, clear, professional book on circadian rhythms.
4. Bünning, E.: "The Physiological Clock," 2d ed., English translation from the German, Springer-Verlag New York Inc., New York, 1967. A superb and fully professional treatment of rhythms in animals and plants.
5. Mohr, H.: Photomorphogenesis, in C. P. Swanson (ed.), "An Introduction to Photobiology," pp. 99–141, Prentice-Hall, Inc., Englewood Cliffs, N.J., 1969. This book is a collection of written versions of symposium lectures given at the Fifth International Congress on Photobiology in 1968. The one by Mohr discusses photoregulation in plants at the level of contemporary research.
6. *Cold Spring Harbor Symp. Quant. Biol.,* vol. 25, 1960. The account of a symposium on biological rhythms.
7. Aschoff, J.: Circadian Rhythms in Man, *Science,* vol. 148, pp. 1427–1432, 1965.

5. Damage and the Repair of Damage by Light

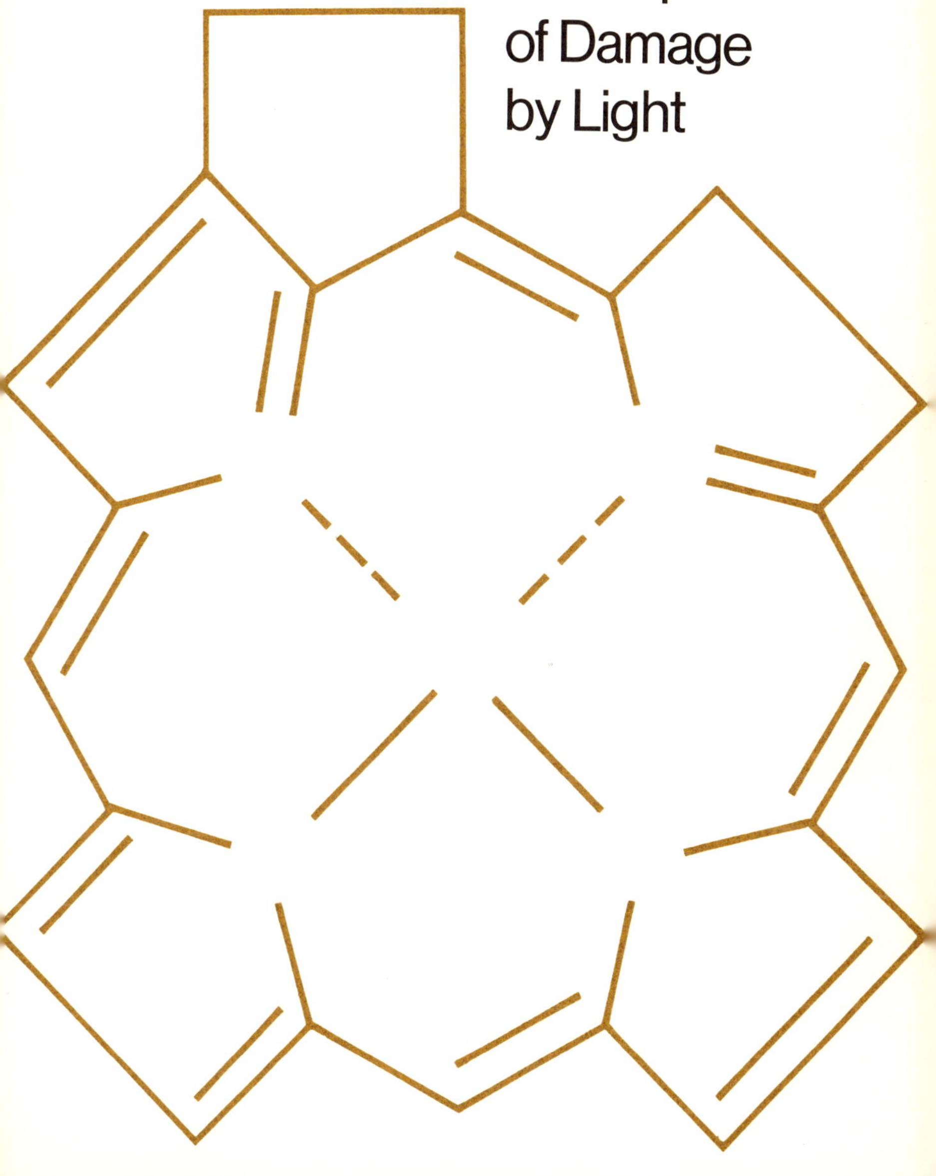

The ability of light to cause biological damage is obvious to anyone receiving a sunburn. The importance of ultraviolet (UV) in causing sunburn was recognized in the nineteenth century, as was the killing of bacteria by UV.

Near the turn of the century N. Finsen made spectacular use of the bactericidal action of UV by curing and curbing the incidence of lupus vulgaris, a tuberculosis of the skin that was common in Northern Europe at that time. In treating this condition Finsen irradiated the skin with both sunlight and artificial UV.

Exposure to sunlight and to UV lamps became a standard part of all tuberculosis therapy. A tradition arose, helped no doubt by the robust appearance of a tanned person, that UV is beneficial to health. This tradition was abetted during the vitamin-consciousness of the 1920s by the discovery that vitamin D, the "sunshine vitamin," is made when precursors (ergosterol and 7-dehydrocholesterol) are exposed to UV. Thus UV prevents rickets, kills germs, and imparts an appearance of vigor.

On the negative side, G. M. Findlay and others (also in the 1920s) learned that mice develop skin cancers after exposure to UV, provided that a UV-absorbing substance (tar or chloroform, for example) has first been applied to the skin. We now correlate human skin cancer, as well as the leathery and wrinkled appearance of aged skin, with exposure to sunlight.

About 1900 O. Raab reported that *Paramecium* cells are killed by light if any of a number of dyes is present in the medium. In humans, an allergic sensitivity to light is induced by a variety of common drugs (especially the tetracycline antibiotics) and by a hereditary disease called porphyria. In this disease the blood carries porphyrins, closed tetrapyrrole molecules, that sensitize toxic photochemical processes.[1]

At this time we recognize that UV and visible radiations, the latter usually with the help of oxygen, cause a wide range of deleterious reactions, especially the inactivation of enzymes and of

[1] An interesting account of the genealogy of porphyria in the royal families of Britain can be found in *The Scientific American;* see Macalpine and Hunter in the Bibliography.

viruses, the killing of bacteria and other cells, mutations of genes and aberrations (gross defects) of chromosomes, and carcinogenesis.

Research has been concentrated especially on the very potent effects of UV absorbed by nucleic acids and proteins, at wavelengths below 300 nm. This has led to a quite detailed understanding of the photochemistry and photobiology of DNA; less so with RNA and protein. But this knowledge is curiously irrelevant to the effects of sunlight at the earth's surface. Ozone, formed by the action of UV on oxygen in the upper atmosphere, absorbs nearly all the sunlight below 300 nm and thereby protects nucleic acids and proteins against UV irradiation. This protective screen is so effective and so important for life on the earth that the formation of ozone from molecular oxygen might be called the most significant reaction in UV photobiology.

Under primordial conditions, before the evolution of oxygen-evolving photosynthesis, the concentration of oxygen in the earth's atmosphere was probably far less and the screening of UV by ozone negligible. The development of life must then have been restricted to deep water, where the UV was screened by inorganic matter. Under these difficult conditions, the evolution of effective ways to repair damaged DNA would have been most advantageous. Contemporary organisms have at least two major ways to undo the effects of UV on DNA. One of these ways is photoreactivation, in which a primary photochemical change is reversed by a light-driven enzymatic process. In the other mechanism, dark repair, the damaged section of DNA is cut out and replaced by a new "good" section. These processes may be less important now than in primordial times, but they are surely of great value in maintaining genetic stability, preventing cancer in higher animals, and reducing the lethal effects of damaged DNA in simpler organisms. Dark repair is effective in curing a variety of lesions in DNA, regardless of their cause.

Oxygen is an important factor in the effects of long-wave UV and visible light on living things. We have learned recently that many of the deleterious photochemical oxidations result from the incorporation of O_2 into bioorganic molecules to form organic peroxides, in highly specific reactions involving excited states of O_2. The excited O_2 is formed from energy absorbed initially by a photosensitizing pigment.

The physical principles (conservation of the symmetry of electronic wave functions, and of electron spin) that dictate the specificity of these reactions of O_2 are useful also for understanding the chemistry of bioluminescence, as we will see in Chap. 6.

THE PHOTOCHEMISTRY AND PHOTOBIOLOGY OF NUCLEIC ACIDS AND PROTEINS

Action spectra for damage to single cells by UV suggest that nucleic acids are the main targets. The maximum effectiveness is usually at 260 to 265 nm, where DNA and RNA have absorption maxima. If proteins were the principal sensitizers, the action spectra would reflect an absorption maximum at about 280 nm. The predominance of nucleic acids appears in action spectra for many effects including killing, delay of cell division, and genetic mutation, measured with bacteria, protozoa, and other single cells. This was first found by F. L. Gates about 1930 for the killing of bacteria and extended in the next two decades by A. Hollaender, A. C. Giese, and many others.

It should be expected that DNA is the most sensitive target for damage in single cells. If a molecule of protein or RNA is destroyed, another can usually be made, but DNA is the primary repository of (genetic) information for the synthesis of everything in the cell including DNA itself. A cell contains only one or a few copies of each specific DNA, so the loss of a single molecule (or part of the molecule) can be irretrievable and fatal for the progeny of that cell. In highly organized forms the continuance of life involves a delicate balance of many chemical and physical factors. A change in a few molecules of either nucleic acid or protein might upset the balance and lead to a self-catalyzing derangement of the whole organism.

5-1. Structure and Photochemistry of DNA and RNA

The discovery that DNA and RNA are major sensitizers of UV damage, together with an appreciation of the importance of these substances in life, has stimulated much study of their photochemistry. To survey this subject let us first note the structures of DNA and RNA, as sketched in Figs. 5-1 and 5-2. In the two-stranded (double helix) structure of DNA each strand is an alternating sequence of orthophosphate and the sugar deoxyribose (ribose in the case of RNA). The purine and pyrimidine bases, shown in Fig. 5-2, are attached to the sugar molecules, and their sequence forms the code that ultimately dictates the sequence of amino acids during the synthesis of specific proteins. The two strands are held together by *hydrogen bonds*[1] between opposed bases. The shapes of the bases and their ability to form hydrogen bonds when juxtaposed dictate a pairwise complementarity: in DNA every guanine is opposite a cytosine, and every thymine is opposite an adenine. In RNA thymine is replaced by uracil.

The absorption maximum near 260 nm in DNA and RNA is a composite of the absorption bands of the individual bases, with only a little modification due to their interactions with their surroundings.

In 1949 R. L. Sinsheimer and R. Hastings showed that both cytosine and uracil lose the 260-nm absorption band when irradiated with UV in aqueous solution. The reaction is reversible; the absorption returns after irradiation. Several years later S. Y. Wang and others showed that this reaction is hydration at the 4,5 double bond, as indicated in Fig. 5-3.

Alternatively two molecules can join as a covalently bonded "cyclobutane" dimer, as shown in Fig. 5-4. The discovery of dimer

[1] A hydrogen bond is an electrostatic attraction between excess positive charge on a hydrogen atom in a molecule and excess negative charge on some other atom, usually oxygen or nitrogen. Its energy ranges from about 3 to 7 kcal/mole. Being such a weak bond it is easily broken and reformed during necessary life processes, but the cooperative action of many such bonds gives the whole DNA molecule the necessary stability.

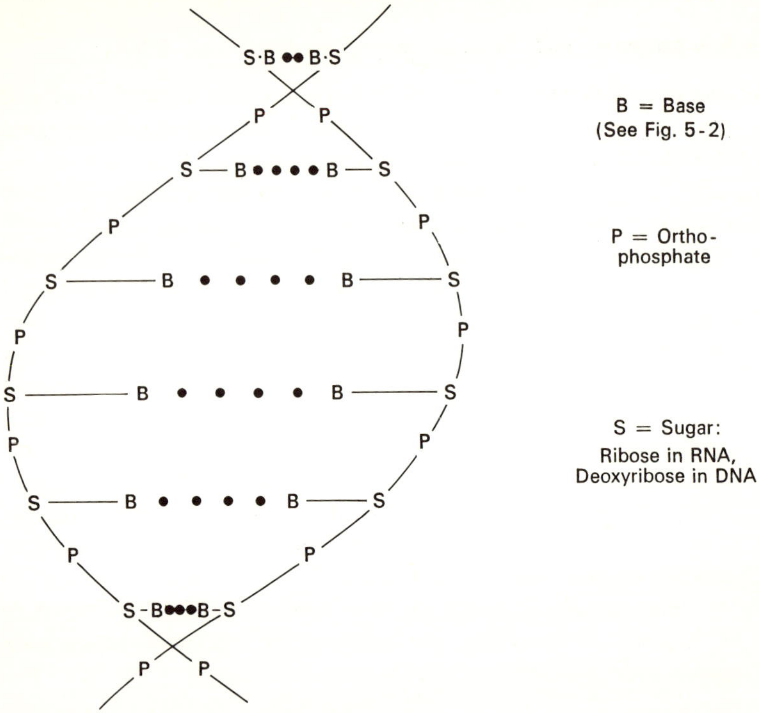

Fig. 5.1 An outline of the two-stranded (double helix) structure of DNA as elucidated by F. H. C. Crick, J. D. Watson, and M. H. F. Wilkins. RNA has a similar structure, with ribose in place of deoxyribose, but it is usually single-stranded.

formation in UV-irradiated pyrimidines was made by R. Beukers and W. Berends about 1960. Irradiation of thymine frozen in ice produces a high yield of dimers, and in contrast to the hydrates of cytosine and uracil the dimers are stable photochemical products. All the pyrimidines can form dimers of the kind shown in Fig. 5-4, either homogeneous, like thymine-thymine, or mixed, like uracil-cytosine. The highest quantum efficiencies are for the formation of thymine dimers.

The behavior of pyrimidines in frozen solution appears to be

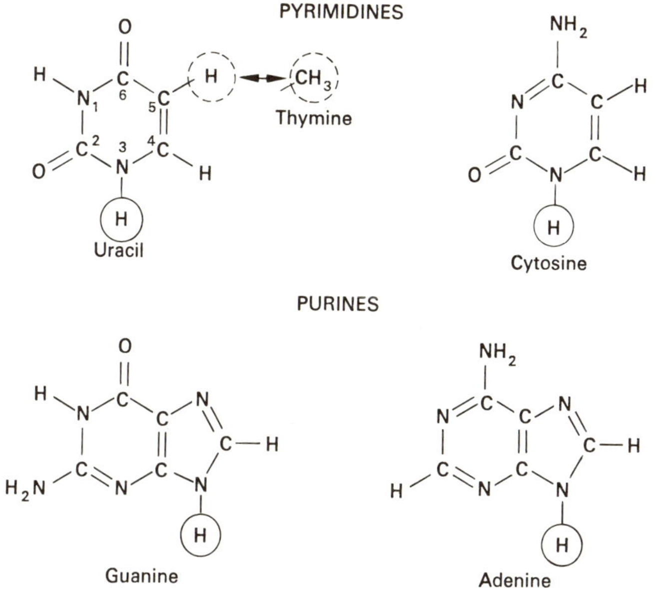

Fig. 5-2. Structures of the bases of DNA and RNA. Both nucleic acids contain the purines guanine and adenine and the pyrimidine cytosine. The pyrimidine thymine is found in DNA, and uracil in RNA. The conventional numbering system in the pyrimidine ring is shown for uracil. These bases are attached to sugar (see Fig. 5-1) by the bonds holding the encircled hydrogen atoms.

pertinent to their behavior when bound in DNA or RNA. Irradiation of DNA causes dimerization of adjacent thymines in one strand [1] with quantum efficiency about 1 percent. This is the predominant result of irradiating DNA; hydrates are far less abundant than thymine dimers, at least as stable products. Other pyrimidine dimers can be made, but with lower quantum efficiency.

[1] The chemical consequences of irradiating DNA or RNA can be determined by analytical techniques that include disruption of hydrogen bonds followed by hydrolysis. A high concentration of urea separates the two strands by breaking hydrogen bonds, and at this stage the appearance of small fragments shows that the strands had been broken. Next the material can be hydrolyzed and assayed for individual bases and dimers. But note (legend of Fig. 5-1) that RNA is usually single-stranded.

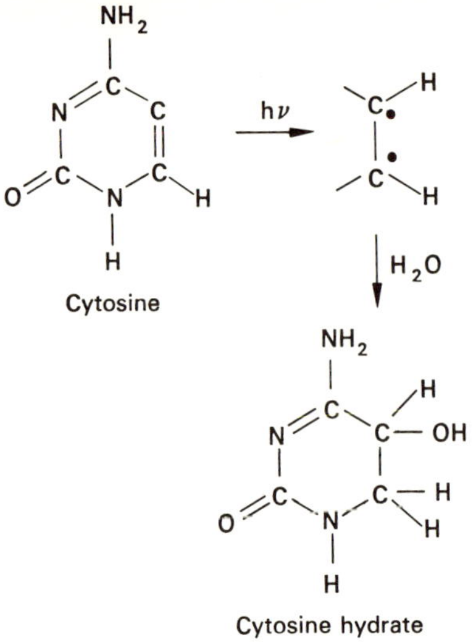

Fig. 5-3. Hydration of cytosine at the 4,5 bond.

In RNA it is not so clear that pyrimidine dimers are the main photoproducts, although the dimers of uracil and cytosine are made in small amounts, at least in frozen solutions of the bases.

The purines are far less reactive than the pyrimidines, perhaps because the extra ring interferes with reactions at the 4,5 bond. They have not been implicated in biological effects of UV.

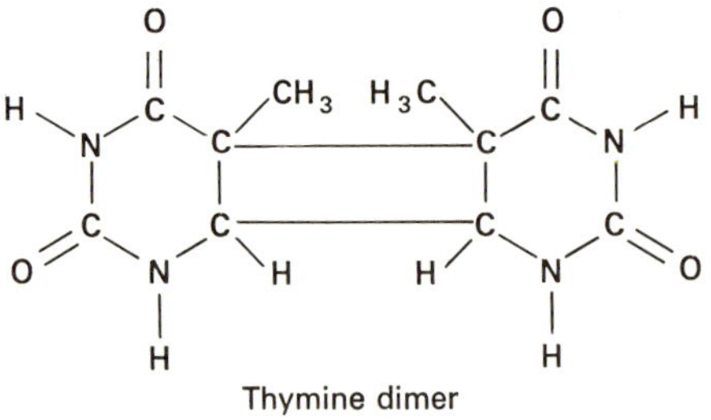

Fig. 5-4. A "cyclobutane" dimer of thymine, formed across the 4,5 atoms. The formation of these dimers accounts for most of the UV-induced damage to DNA.

The dimerization of pyrimidines is photoreversible: the dimers can absorb UV, and can be restored photochemically to monomers. The forward and reverse reactions, formation and splitting of dimers, become balanced in a photostationary state, as with the two forms of phytochrome (recall Secs. 4-6 and 4-7). The balance between dimer and monomers is set by the relative values of $\varepsilon\phi$ (extinction coefficient times quantum efficiency) for the forward and reverse processes. For thymine in ice, ϕ is about 0.1 for dimerization and close to 1.0 for the reverse reaction. The corresponding values of $\varepsilon\phi$ for monomer and dimer are sketched as functions of wavelength in Fig. 5-5. These curves predict the action spectra for formation and splitting of dimers, and the predictions are in good agreement with measurements by R. B. Setlow and others.

We see from Fig. 5-5 that dimers should predominate after irradiation at 280 nm, whereas 240-nm irradiation should shift the balance in favor of monomers. This can explain the odd finding that biological damage inflicted by 260- to 280-nm light can sometimes be alleviated by a subsequent dose of 240-nm light.

In summary, the main consequence of exposing DNA to UV is the formation of thymine dimers. Among lesser effects we can list the formation of other dimers and of hydrated pyrimidines, the breaking of a strand, and the formation of cross links between the two strands. A local disruption of hydrogen bonds (denaturation) might follow any of these chemical changes. In RNA, lacking thymine, there is no single outstanding known photoproduct, and the photochemical picture is not very clear. We shall see that these chemical facts, especially those pertaining to thymine dimers, are reflected convincingly in biological effects of UV.

5-2. Photobiology of Nucleic Acids and Proteins

If photochemical experiments are made with nucleic acids or proteins that have biological activity, then the chemical and biological consequences of UV irradiation can be compared. This can be done at different levels of simplicity. If whole cells are irradiated, the photochemical products interact with a variety of meta-

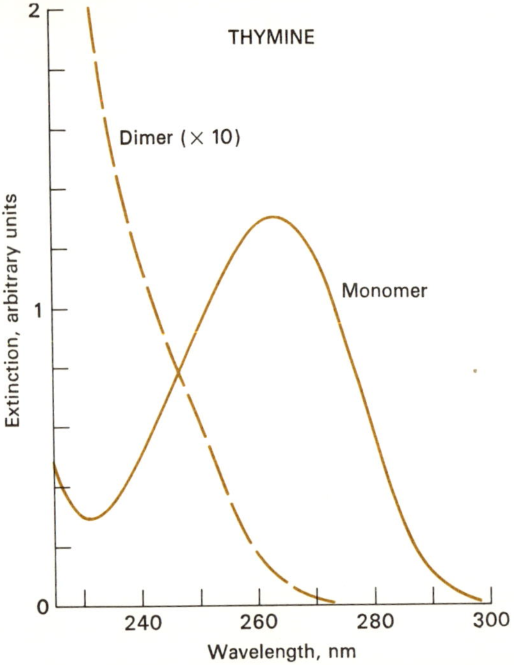

Fig. 5-5. Absorption spectra of thymine monomer and dimer. The molar extinction of the dimer has been exaggerated tenfold. The two curves then represent the rates of formation and splitting of dimer in frozen solution, because the quantum efficiency for dedimerization (when light is absorbed by the dimer) is ten times greater than that for dimerization (light absorbed by monomer). These curves are thus plots of the product of the extinction coefficient and the quantum efficiency. The ratio of their heights at any wavelength predicts the relative concentrations of dimer and monomer in a photostationary state established at that wavelength; compare with Fig. 4-3 (center).

bolic systems (most significantly, with enzymes that repair some of the damage) before the result can be seen. A great simplification is introduced when we study viruses, which are generally made of either DNA (in the case of most viruses that infect bacteria) or RNA (plant viruses) plus some protein. The infectivity of the virus can be related to photochemical changes of its components, and enzyme systems derived from other sources can be added or withheld at will. Still simpler objects of study are purified enzymes and nucleic acids that have biological activity. A great deal has been learned about UV photobiology through experiments with one such substance, transforming DNA.

If DNA is extracted from one type of bacterium and added to a culture of another closely related strain, the genetic information carried by the DNA can sometimes be transferred. The recipient in this transduction acquires some genetically controlled traits of the donor. As an example, resistance to streptomycin can be transferred from a resistant to a sensitive strain of *Hemophilus influenzae*. In this case the biological activity of the transforming DNA is measured by its ability to confer streptomycin resistance on the recipient strain.

We owe much of our knowledge of the photobiology of DNA to studies, especially by C. S. Rupert, R. B. Setlow and J. K. Setlow, and their associates, made with transforming DNA from *H. influenzae*. To appreciate this work we must be aware of two biological mechanisms, described more fully in Secs. 5-5 and 5-6, for repairing damaged DNA. The first is *dark repair:* an enzyme system, on recognizing a defective spot on a DNA molecule, cuts out the "bad" section (one strand only). The undamaged complementary strand then acts as a template for the synthesis of a new "good" section to replace the discarded piece.[1] Operation of this mechanism is signaled by accumulation of the discarded fragments that contain photochemically altered components, such as

[1] The ability of one strand to direct the replacement of a defective region in the other strand shows the great biological value of the two-stranded structure in DNA. Single-stranded DNA could perform most of the functions of the usual double-stranded form, but would be unable to govern the repair of its own lesions.

pyrimidine dimers. The dark-repair enzyme system is found fairly universally in cells and higher organisms, but not in viruses (which must rely on the enzymes of their hosts).

The second major mechanism for restoring DNA is photoreactivation, a specific light-dependent means of splitting pyrimidine dimers and thus undoing the damage of dimerization. This is distinct from the photoreversal of dimerization, described earlier, caused by UV near 240 nm absorbed directly by the dimers. Photoreactivation is effective with long-wave UV, violet, and blue light, and involves an enzyme found in most living things. The enzyme binds to DNA that contains pyrimidine dimers. Light absorbed by this enzyme-DNA complex causes the dimers to be split. The enzyme and the DNA then separate. The results of photoreactivation thus differ from those of dark repair in two principal ways: (1) Photoreactivation repairs only pyrimidine dimers as far as we know, whereas dark repair can heal other lesions in DNA. (2) Photoreactivation restores the DNA by converting dimers back to monomers without performing new syntheses or leaving byproducts.

Transforming DNA can be irradiated with UV, then exposed (or not) to one of these repair systems, and finally analyzed for biological activity and residual photochemical alterations. In this way the Setlows showed that after irradiation near 280 nm, the application of 240-nm UV is one way to reduce the amount of thymine dimer and restore biological activity, and photoreactivation is another way. To the extent that the dedimerization and biological restoration has been effected by one of these means, less remains to be done by the other. These experiments showed that most of the UV damage to transforming DNA can be attributed to dimer formation.

In cells and DNA-containing viruses, as well as with transforming DNA, a correlation could be drawn, after irradiation and photoreactivation, between biological damage and the formation of pyrimidine dimers. With bacterial cells (*Escherichia coli*) the correlation could be extended to mutant strains having abnormally high or low sensitivity to UV. The UV-resistant strains have highly active dark-repair systems, evidenced by greater release of thymine dimers during a period of repair after irradiation. The sensitive mutants correspondingly show abnormally little release of dimers after exposure to UV.

The average dose of UV (near 260 nm) for 50 percent survival in a culture of "normal" *E. coli* cells is about 10,000 ergs/cm^2, enough to make hundreds of thymine dimers in each cell. In a sensitive strain, deficient in dark repair, the corresponding dose is only 100 ergs/cm^2. This suggests that in the normal strain about 99 percent of the damage is repaired and that a few unrepaired thymine dimers in one cell are likely to be fatal. The fact that the lethal effect of UV is reduced twofold to tenfold by photoreactivation indicates that in bacteria, as well as in transforming DNA, most of the UV damage is associated with formation of pyrimidine dimers.

After highly effective photoreactivation, less common types of damage to DNA might become relatively more significant in the residual damage by UV. These are not well characterized, but could include hydration of pyrimidines, single-strand breaks, and local disarrangement of hydrogen bonds. Also, when spores are irradiated, the damage to their nucleic acids is not photoreactivable, and no pyrimidine dimers are found. The "spore DNA photoproduct" is unknown.

In bacteria with unusually effective mechanisms for repairing DNA (an example aside from *E. coli* mutants is *Micrococcus radiodurans*), the reactions of proteins also acquire relatively more significance. This can be seen in the action spectrum for killing, which may show equal prominences at 285 nm (protein) and 260 nm (nucleic acids).

According to A. D. McLaren the action of UV on proteins, as measured by the inactivation of enzymes, can be accounted for roughly as the sum of the effects on the individual amino acids of which the proteins are composed. For most of these amino acids, UV causes a poorly characterized assortment of reactions: deaminations (replacement of $-NH_2$ by $-H$ or $-OH$), decarboxylations (replacement of $-COOH$ by $-H$ or $-OH$), ring breaks, and so forth. The most sensitive targets are the aromatic amino acids, especially tryptophane, simply because they have the greatest ability to absorb UV. Quantum efficiencies for reactions of these amino acids are of the order of 10^{-3} to 10^{-2}. But cystine, with a tenfold lower extinction coefficient than tryptophane in the neighborhood of 280 nm, has a special importance. Its disulfide bond (—S—S—) is broken with a quantum efficiency of about 0.1, and this bond is important in maintaining the shape (tertiary structure) of most enzyme proteins. The sensitivity of enzymes to inactivation by UV shows a close correlation with the cystine content.

K. C. Smith and others have studied the possibility that the formation of cross links between DNA and protein can be important in UV damage to cells. The cysteine residues in proteins (with —SH groups) can be shown to combine with pyrimidines.

These processes seem relatively minor in laboratory experiments showing UV damage to single cells, being entirely overshadowed by dimerization in DNA as long as the UV is of wavelengths below 300 nm. However, the "lesser" reactions may have far more relative importance in the natural setting at the earth's surface, where long-wave UV and visible light predominate.

Viruses that contain RNA are simple objects of study, but the results of UV irradiation have not shown the simple unity seen with DNA, where one photoproduct is of overriding importance. Pyrimidine dimers can be made in UV-irradiated RNA; this is suggested by the fact that some of the damage to plant viruses can be photoreactivated. However, the extent of photoreactivation is variable and is negligible in the case of RNA from bacterial viruses and cells. The case of tobacco mosaic virus exemplifies the situation. If the intact virus, RNA plus a coat of protein, is irradiated with UV, damage is shown by the loss of the ability to infect tobacco leaves. The viral RNA, with the protein coat removed, is also infective. This RNA is about five times as sensitive to inactivation by UV as the intact virus. About half of the damage to the protein-free RNA can be photoreactivated by an enzyme in bean leaves. In contrast, the damage inflicted by UV on the intact virus cannot be photoreactivated, even if the protein coat is removed before the attempt at photoreactivation. Perhaps the presence of the protein constrains the shape of the RNA so as to prevent the formation of dimers, resulting in less sensitivity to UV and no photoreactivation.

RECOVERY FROM DAMAGE BY ULTRAVIOLET

5-3. Primitive Earth Conditions

As mentioned earlier, the development of life in primordial times (about a billion years ago) was probably restricted by the high intensity of shortwave UV at the earth's surface.[1] When primitive algae invented oxygen-evolving photosynthesis, the release of oxygen was probably disastrous for most things living at that time. Organisms either succumbed to oxygen poisoning,[2] or found anaerobic refuges, or learned to deal with this new poison and even to use it for life processes (development of the respiratory way of life).

The concentration of oxygen eventually rose to such levels that the screening of UV by ozone allowed life to proliferate at the surfaces of the seas. This led to an explosive spurt of evolution, and what began as an "aerobic catastrophe" became a blessing as measured by the abundance and diversity of life.

The present concentration of ozone in the upper atmosphere is enough to reduce the level of UV at the earth's surface by about 200 times at 300 nm, by about 10^7-fold at 290 nm, and to utterly negligible levels at shorter wavelengths. Mechanisms for the repair of UV damage are not as crucial for life as they probably once were, but such mechanisms continue to be effective and widespread among living things.

[1] For discussions of primeval biochemical and biological evolution see the works by Oparin and by Berkner and Marshall listed in the Bibliography.
[2] The mechanism of "oxygen poisoning" is not clear, but many contemporary types of bacteria are killed or are at least prevented from multiplying when exposed to oxygen at concentrations down to about $\frac{1}{1,000}$ atmospheric.

5-4. Analysis of Relations between Dose and Effect in Terms of Target Theory

An important element in the study of radiation biology has been the analysis of relations between the dose of radiation and the biological effect. This has provided information, for example, about the number of photochemical lesions that must be inflicted in order to kill a cell. The dose-effect relation also gives evidence for the existence of repair mechanisms, and for relationships between different processes in damage and repair.

Consider an effect such as the killing of a cell or the inactivation of a molecule, and ask the question "How many independent lesions are needed to produce the effect?"

In other words, how many independent targets for radiation are there (in one cell, molecule, or other entity), such that *every* target must be hit in order for the effect to occur? As examples, in a haploid cell it might be enough to hit the one and only gene of a certain kind, but in a diploid cell, where there are two genes of each general kind, it might be necessary to hit both genes of a certain type. Again, consider a bacterium of such a type that in liquid culture the cells tend to stick together in pairs. If a little of the cell suspension is spread over a culture plate, each cell pair will multiply and form a visible colony on the surface of the plate. If one cell of a pair is killed, the other cell will still give rise to a colony. To prevent the formation of a colony it is necessary to kill both cells of a pair. For this organism the prevention of colony formation (the usual criterion of killing) presents two independent targets for each operational entity.

We will formulate first the probability that any one target is hit after a certain dose of radiation and then the probability that all the targets in one entity are hit.

Let the average number of quanta hitting one target be denoted $\bar{n}$. This will be proportional to the dose D of radiation, in whatever units are used:

$$\bar{n} = kD$$

The Poisson distribution (see Sec. 3-4) gives the probability $P(n)$ that a target actually receives n hits, if the average number is $\bar{n}$:

$$P(n) = \frac{(\bar{n})^n e^{-n}}{n!} \qquad (3\text{-}2), (5\text{-}1)$$

The probability that a certain target is missed is then

$$P(0) = e^{-n} = e^{-kD} \qquad (5\text{-}2)$$

and the probability that a target gets *at least* one hit is

$$p = 1 - P(0) = 1 - e^{-kD} \qquad (5\text{-}3)$$

Now that we know the probability of hitting any one target with a dose D, we must compute the probability that *all* the independent targets in one entity (cell, molecule, etc.) are hit. Let the target multiplicity be m, and consider that "shooting" at each of the m targets represents an independent trial, with a probability of success (hit) given by P in Eq. (5-3). The probability of hitting x out of m targets is given by the binomial distribution

$$Pr(x) = \left[\frac{m!}{x!(m-x)!}\right] p^x(1-p)^{m-x} \qquad (5\text{-}4)$$

Now the probability of hitting all m targets is

$$Pr(m) = p^m(1-p)^0 = p^m$$

or, from Eq. (5-3),

$$Pr(m) = (1 - e^{-kD})^m \qquad (5\text{-}5)$$

Finally, the probability of failing to hit all m targets after a dose D

(the probability of survival) is

$$Pr(\text{survival}) = 1 - Pr(m) \tag{5-6}$$

Equating this survival probability to the fraction of cells that are expected to survive a dose *D*, we have

$$\frac{N}{N_0} = 1 - (1 - e^{-kD})^m \tag{5-7}$$

where N_0 is the initial number of cells (or other entities) and *N* is the number that survive.

If a cell can be killed by hitting just one target ($m = 1$), we have the "one-hit" survival relationship [1]

$$\frac{N}{N_0} = e^{-kD} \tag{5-8}$$

Equation (5-8) can be expressed and plotted in logarithmic form,

$$\log \frac{N}{N_0} = -kD \log e = -k'D \tag{5-9}$$

giving a straight line for $\log(N/N_0)$ versus *D*, as shown in Fig. 5-6.

Equation (5-7), for $m > 1$, takes on simpler forms at very high and very low doses. If $kD << 1$, $e^{-kD} \approx 1 - kD$ and Eq. (5-7) becomes

[1] This is like the expression derived in an entirely different way (in Sec. 3-3, Volume 1) for the attenuation of a light beam in an absorbing medium:

$$\frac{I_x}{I_o} = e^{-aCx} \tag{3-4}$$

The analogy is complete if we interchange the roles of molecules and quanta, and think of the chance that a quantum survives being inactivated (absorbed) by a certain "dose" of light-absorbing material *Cx*.

$$\frac{N}{N_0} \approx 1 - (kD)^m$$

and in logarithmic form, to base 10,

$$\log \frac{N}{N_0} \approx -\frac{1}{2.3}(kD)^m \qquad (5\text{-}10)$$

since $\log(1 + x) \approx x/2.3$ if $x << 1$. In this application x is defined as $-(kD)^m$. Equation (5-10) plots as a curve whose slope approaches zero at the origin (see Fig. 5-6, curve for $m > 1$).

For high doses, where $kD >> 1$, Eq. (5-7) reduces to

$$\frac{N}{N_0} \approx me^{-kD}$$

or

$$\log \frac{N}{mN_0} \approx -k'D \qquad (5\text{-}11)$$

This is identical to Eq. (5-9) for the one-hit case except that it extrapolates to $N = mN_0$ rather than $N = N_0$ at zero dose, as shown in Fig. 5-6. Thus the number of targets m can be learned simply by extrapolating the survival curve in this way.

The success of this treatment was evident in many studies of survival under ionizing radiations. For example, yeast cells can be produced in the haploid and the diploid condition, and these give one-hit and two-hit survival curves respectively under irradiation with x-rays. For UV inactivation of enzymes, transforming DNA, and viruses the survival curves are generally one-hit, but with bacteria the curves often do not resemble any of those predicted by the target theory. Often the survival curve is a sort of hybrid between the one-hit and many-hit curves, as sketched as a dotted line in Fig. 5-6. This has been interpreted as signifying the occurrence of repair processes, superimposed on a one-hit situation. At low doses the repair diminishes the effect of the UV and

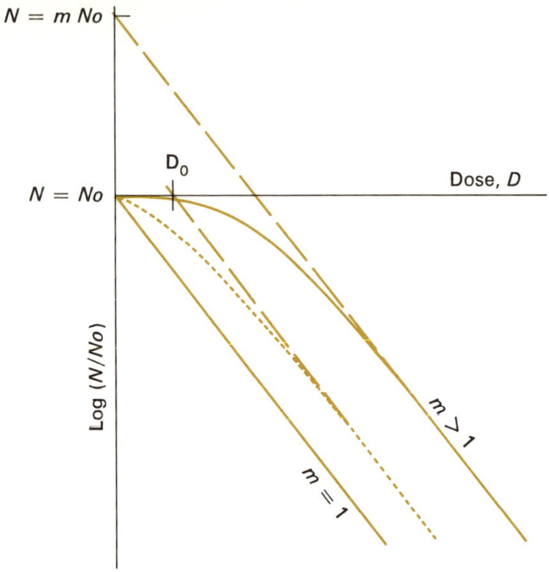

Fig. 5-6. Theoretical plots of survival in a population of entities as a function of the dose D of radiation. The plot is semilogarithmic, log (N/N_0) versus D where N/N_0 is the fraction surviving. The quantity m represents the number of independent targets in each entity, all of which must be hit in order to kill the entity. The dotted curve shows the expected consequence of a repair process operating in a system for which $m = 1$. If the straight part of the dotted curve is extrapolated back to the horizontal axis, the intercept D_0 shows how large a dose can be nullified by repair.

gives a higher surviving fraction. At higher doses the capability of the repair system becomes saturated, and the survival curve becomes linear. If this linear part is extrapolated back to the *horizontal* axis, one gains an idea of how large a dose (D_0) can be nullified by the repair process.

5-5. Dark Repair of DNA

The shapes of survival curves provided one line of evidence that bacteria possess a mechanism for repairing UV damage. A normal strain of *E. coli* may show a linear or one-hit survival curve, with about 10^4 ergs/cm^2 for $N = \frac{1}{2}N_0$. For a resistant strain, the slope of the curve for high doses may be like that of the sensitive strain, but at low doses the curve is shaped like the dotted curve of Fig. 5-6. In the resistant strain nearly all the damage at low doses is repaired.

Probably the earliest evidence for repair mechanisms was the "liquid holding recovery" phenomenon reported by A. Hollaender and others in the 1930s. If spores or bacteria are irradiated, their eventual survival is enhanced if they are not immediately challenged to grow and divide. If the cells are held in water, away from nutrients, for a few hours after irradiation, their repair system seems to have a chance to function before the DNA is called upon to give genetic information. As a result the survival is greater, and the survival curve may change from the "one-hit" form to the form of the dotted curve in Fig. 5-6. During the recovery phase (cells withheld from nutrients) there is a release of thymine dimers correlated with the degree of recovery, as shown by R. B. Setlow and W. L. Carrier with *E. coli*.

Dark repair of DNA has been studied extensively by P. C. Hanawalt and R. H. Haynes, among others. The process is roughly as follows: A local defect in DNA is "recognized" by an enzyme system which breaks the defective strand near the lesion. A piece containing about twenty bases is peeled away, cut off, and ejected. The undamaged complementary strand directs an enzymatic synthesis of a replacement for the discarded piece.

This process is of great general importance. It works regardless of what caused the lesion (UV, chemicals, etc.). It opens the door to the possibility that we might develop a kind of "genetic engineering" which could have formidable consequences.

Dark repair appears to keep the frequency of cancers down to a "tolerable" level. In the rare skin disease xeroderma pigmentosum the skin cells lack the dark-repair mechanism and have a great tendency to become malignant, especially in the sun.

Although the repair process surely reduces the number of defects in DNA, it is possible that errors during repair may leave residual mutations due to an improper sequence of bases.

The Setlows and Carrier have performed an experiment in which the dark-repair enzyme system extracted from a bacterium (*Micrococcus lysodeikticus*) was added to UV-irradiated transforming DNA. They found recovery of biological activity, and during the repair period they also detected single-strand breaks and corresponding numbers of thymine dimers released, consistent with the foregoing view of the repair process. The one-to-one correspondence between the strand breaks (made in order to effect repair) and the release of dimers showed again that most of the lesions were due to thymine dimerization.

Because repair enzymes from cells can be tested with viruses or transforming DNA, it is possible to survey the world of life and to show that most living things possess the system for repairing DNA. Human cells have highly effective dark-repair enzymes.

5-6. Photoreactivation: Repair of DNA and RNA

A. Kelner discovered photoreactivation in 1949 while trying to quantitate the dose-effect relationships for the killing of bacteria (*Streptomyces*) by x-rays and UV. There was no problem with x-rays, but with UV there were odd variations that could finally be traced to the restoring action of visible light. R. Dulbecco then made the important discovery that UV damage to bacterial viruses could be alleviated by visible light, but only in the presence of the host bacteria. The viral DNA could not effect photoreactivation by itself.

Soon the observation of photoreactivation was extended to many systems: bacteria, fungi, algae, and higher plants and animals. The phenomena that could be caused by UV and relieved by photoreactivation were found to include not only killing but also mutation, inhibition of chloroplast development, inhibition of motility in algae, and even retardation of the phase of a circadian clock. A universal correlation developed, after a time, to the effect that photoreactivation can occur only in connection with damage to nucleic acids.

Probably spurred by Dulbecco's finding that viral DNA can be photoreactivated in the presence of host cells, C. S. Rupert and others attempted to induce photoreactivation of transforming DNA with the help of extracts from cells. The first well-defined success was reported in 1958 by S. H. Goodgal, C. S. Rupert, and R. M. Herriott. These workers mixed an enzyme extracted from *E. coli* with transforming DNA from *H. influenzae,* after the DNA had been exposed to UV. Illumination of the mixture with long-wave UV caused a partial restoration of the transforming activity of the DNA. A similar experiment was then made by Rupert using a photoreactivating enzyme obtained from yeast. He and his colleagues showed that the enzyme binds to UV-irradiated DNA, and is released in consequence of the photochemical act that restores the DNA. The Setlows and others at Oak Ridge National Laboratory participated in showing that photoreactivating enzyme has specific affinity for nucleic acids that contain pyrimidine dimers. As far as we know, photoreactivation is effective only in the splitting of these dimers in DNA and RNA (recall the evidence cited in Sec. 5-2). In materials that have been irradiated with UV and then exposed to photoreactivating enzyme and light, the extent of photoreactivation is generally correlated with a reduction in the content of pyrimidine dimers. And in systems which cannot be photoreactivated, such as dry spores or intact tobacco mosaic virus, UV damage does not involve identifiable dimer formation.

Photoreactivating enzyme has now been found in nearly all cellular and multicellular forms of life, with the notable exception of most mammals. The enzyme from yeast has been purified several thousandfold by A. Muhammed.

Action spectra for photoreactivation generally show activity somewhere between about 300 and 500 nm, but the spectra differ widely for photoreactivating enzyme from various sources. Thus for photoreactivation of *E. coli* (killing) the action spectrum shows three maxima: at about 325, 350, and 380 nm. For *Streptomyces griseus* there is a peak at 435 nm and lesser shoulders at shorter wavelengths. If a single photoreactivating enzyme is tested with different DNAs, including synthetic ones (with these, photoreactivation means the splitting of dimers), the action spectra show both similarities and slight differences.

These facts are reminiscent of the absorption spectra of different rod and cone pigments, produced by the union of different retinals with different opsins. The analogy to vision, promoted by Rupert among others, provides a very simple picture of photoreactivation: it is nothing more than the photochemical splitting of pyrimidine dimers, like that caused by 240-nm light absorbed by the dimer in the absence of photoreactivating enzyme. The only difference is that the absorption maximum of the dimer has been shifted, because of binding to the enzyme, from below 210 nm to somewhere in the 300- to 500-nm region. In this view the enzyme has no chromophore of its own; the only chromophore is the enzyme-bound dimer.

Two things militate against this economical picture. First, a shift of the absorption maximum from less than 210 nm to as far as 435 nm would be unprecedentedly large. Even with iodopsin the wavelength does not change by as much as a factor of 2. Second, the complex spectra such as the triple-banded one in *E. coli* would be hard to explain.

Whether the chromophore is in the enzyme only or involves the cooperation of pyrimidine dimers, it is puzzling that no one has yet been able to see an absorption band corresponding to the action spectrum for photoreactivation, even in highly purified photoreactivating enzyme bound to UV-irradiated DNA.

DAMAGING EFFECTS OF LONG-WAVE ULTRAVIOLET AND VISIBLE LIGHT

5-7. A Variety of Biological Effects

Long-wave UV and visible light, the latter aided by oxygen, can kill bacteria, but far less effectively than UV in the 200- to 300-nm region. Between 350 and 500 nm the number of quanta needed for 50 percent killing of *E. coli* is 10^3 to 10^4 times greater than at the 260-nm peak. The long-wave killing is not attended by the induction of mutations in the surviving cells, as is the effect of shortwave UV. The survival curves invariably show a high multiplicity of targets with long-wave killing, even if the response to shortwave is one-hit.

One special exception to these remarks is found when substances absorb long-wave UV and transfer the excitation energy to nucleic acids. A clear laboratory example is the case where dimerization of thymine in DNA is sensitized by 300- to 400-nm light absorbed by acetone or acetophenone. We cannot conceive that energy transfer is by the obvious singlet excited states. A state corresponding to absorption at wavelengths greater than 300 nm has perhaps 20 kcal/mole less energy than one represented by the 260-nm absorption band of a nucleic acid. A process of energy transfer in which this much energy must be gained from the surroundings would have vanishingly small probability. On the other hand we see no evidence for chemical participation by the sensitizer. This leaves the likely possibility that energy is transferred from the sensitizer to the thymine in DNA by way of triplet excited states:

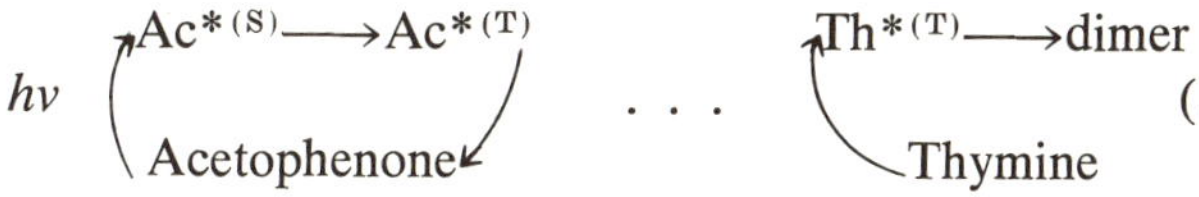

(5-12)

where $Ac^{*(S)}$ and $Ac^{*(T)}$ are singlet and triplet excited states of the sensitizer acetophenone. The energy of $Ac^{*(T)}$ is greater than that of $Th^{*(T)}$, the triplet state of thymine.

In living cells, substances such as flavins, pyridine nucleotides (NAD), and quinones might play the role of long-wave UV sensitizers.

Reactions of this kind provide the strongest evidence that photochemistry of nucleic acids goes by way of triplet excited states. Other evidence is based on fluorescence as a measure of singlet excited states and phosphorescence as a measure of triplet states (see Sec. 2-11, Volume 1). If fluorescent or phosphorescent molecules are added to DNA and the DNA is excited with UV, the fluorescence or phosphorescence shows that the dye has taken energy (as singlet or triplet excitation, respectively) from the DNA. This drain of energy can compete with the photochemical formation of thymine dimers. Phosphorescent dyes often compete strongly, whereas fluorescent ones usually do not. This indicates that energy is transferred through the DNA molecule, to thymine or to the dye, by way of triplet states.

Oxygen interacts strongly with molecules in triplet excited states, as we shall see. The presence of oxygen therefore inhibits reactions of the foregoing type by taking away the excitation energy. When this happens, a new class of reactions, featuring the chemical participation of oxygen, becomes important. In this category there are two major possibilities that will be discussed in Secs. 5-9 and 5-10. First, O_2 may act as electron acceptor in a dye-sensitized electron transfer reaction that causes oxidation of vital substances. Second, O_2 may be induced photochemically to combine with molecules so as to form organic peroxides. Even if the oxygenated molecule was not itself vital, it could become dangerous as a peroxide. Toxic lipid peroxides can be found in tissues that have been exposed to light in the presence of oxygen.

Some of the better-known consequences of long-wave UV and visible irradiation will now be described.

An interesting laboratory curiosity is *photoprotection:* a reaction induced by long-wave UV that alleviates the effects of other radiations such as shortwave UV. This is distinct from photoreactivation in that it works when the "protecting" light is given before the damaging radiation. Photoprotection may be a variation of *liquid-holding recovery.* The long-wave UV, perhaps absorbed by quinones that are involved in respiration, slows the cell's metabolism. This gives the dark-repair systems a better chance to work before the damaged cell exposes itself by attempting to use its DNA.

The delay or prevention of cell division is one of the more sensitive responses of cells to light, studied particularly by A. C. Giese and also by R. H. Haynes. Haynes has implicated a potent but unknown chemical, formed by UV, that destroys the mitotic spindle apparatus.

In photosynthetic tissues the Chl can sensitize harmful reactions that depend on oxygen. These reactions, manifested by killing and by destruction of the Chl, are prevented by the carotenoid pigments that are universally present in photosynthetic organisms in nature. Deleterious photooxidations sensitized by Chl (and BChl) were discovered in laboratory strains of photosynthetic bacteria and algae that lack colored carotenoids. It was then found that carotenoids can also protect nonphotosynthetic cells (bacteria, molds, and even animal cells) against damage by blue light in the presence of oxygen.

Carotenoids are known to quench triplet excited states in many dyes including Chl, probably by a transfer of the excitation energy to the carotenoid. This can account for the ability of carotenoids to inhibit many photochemical processes in vitro, in which the chemistry is initiated by the triplet state of the sensitizing dye. Electron transfer reactions sensitized by Chl in vitro, with and without the participation of oxygen, are inhibited strongly by carotenoids that have nine or more conjugated double bonds. The same carotenoids are effective in quenching the triplet excited state of Chl in vitro, and in preventing damage sensitized by Chl in vivo. Carotenoids with fewer than nine conjugated double bonds are relatively ineffective. It may be that with nine or more conjugated double bonds (but not fewer) the triplet state of the carotenoid is lower in energy than that of the Chl, facilitating the transfer of triplet excitation from the Chl to the carotenoid. In the reactions with oxygen, the carotenoids might also protect by serving as harmless substrates in a cycle of oxidation and reduction, or by quenching the excited states of oxygen (see Sec. 5-10).

Toxic light reactions can be sensitized in animals by pigments of natural origin and by light-absorbing substances that are ingested. In the first category we have the disease porphyria mentioned earlier. Second, a variety of common drugs are phototoxic, and can induce allergic reactions in persons exposed to sunlight. These drugs include steroids (many hormones), sulfanilamide, chlorpromazine, and the tetracycline antibiotics.

Finally there are the familiar responses of human skin to sunlight. These are of sufficient special interest to merit a separate section.

5-8. The Photophysiology of Human Skin

The outermost layer of human skin, the *stratum corneum,* is made of dead skin cells. It is about 25 microns thick over most of the body but can become much thicker. Beneath this is the other layer of the *epidermis,* the *malpighian layer,* which contains living skin cells. The skin cells, called *keratinocytes,* are made by division, differentiation, and outward migration of cells derived from a basal cell layer, still within the epidermis. A special class of cells called *melanocytes* store the brown-black pigment melanin and can inject some of this pigment into the keratinocytes. The malpighian layer is about 25 microns thick.

Under the epidermis is the *dermis,* a layer made up mainly of connective tissue and blood vessels. The dermis also contains mast cells and lysosomes; these bodies, when injured, release vasodilator substances and protein digesting enzymes.

Exposure of skin to UV causes the familiar signs of sunburn: a redness due to vasodilation, and pain. This inflammation reaction, which develops in about 2 hours, may be triggered by the rupture of mast cells and lysosomes; the process is not well understood. A local concentration of histamine is often associated with inflammations, but this correlation is not convincing in sunburn.

The action spectrum for sunburn shows a sharp maximum at 300 nm and a minimum at 280 nm. The minimum is due to the screening action of protein and nucleic acid in the stratum corneum and the malpighian layer; these layers of the epidermis have an OD at 280 nm of about 1 for a thickness of 25 microns. Because of this intense screening at shorter wavelengths, the peak at 300 nm in the sunburn action spectrum is unusually sharp. This peak stands at the very limit of the wavelengths in sunlight that reach the earth's surface. At 300 nm, a skin thickness of about 50 microns has an OD of 1, corresponding to 10 percent transmission.

The inflammation of moderate sunburn subsides in about two days, while the protective mechanisms of skin growth and tanning are getting under way. There is more frequent cell division in the basal layer of skin cells, producing more cells and subsequently a thicker stratum corneum. This increased activity reaches a peak in about one week and subsides in the second week. The resulting thicker stratum corneum, perhaps 80 instead of 25 microns, may last for weeks. While present, the extra thickness can easily reduce by a factor of 10 the intensity of harmful light reaching the lower epidermis and the dermis.

The main tanning process also develops after about two days and reaches a peak after about one week. The melanocytes generate more granules of melanin and disperse more of this pigment into the keratinocytes. The effect is like that of opening a bottle of ink in the bathtub (sieve effect; Volume 1, page 83). With greater dispersal of the pigment, from melanocytes into all of the skin cells, the color is more apparent, and the absorption of light becomes much more effective.

These two responses, growth of the stratum corneum and tanning, have roughly equal value in preventing further sunburn. The action spectra for these responses are the same as that for sunburn.

There is another tanning process which happens immediately upon exposure to sunlight, which requires oxygen, and which has a different action spectrum with a broad band centered at about 350 nm. This process is the photooxidative formation of melanin from a precursor substance. This prompt tanning has less prolonged protective value, but it can erect a quick screen to reduce sunburn.

The hazard of sunlight as a carcinogenic agent, and its ability to give skin an aged appearance, need not be belabored further.

5-9. Photochemical Electron Transfer and the Participation of Oxygen

Consider the photochemical electron transfer shown in Fig. 1-1, involving a sensitizer S, an electron donor D, and an acceptor A:

$$A + S + D \xrightarrow{h\nu} A + S^* + D \left[\begin{array}{l} \rightarrow A^- + S^+ + D \\ \rightarrow A + S^- + D^+ \end{array} \right] \rightarrow A^- + S + D^+ \qquad (5\text{-}13)$$

Photochemical oxidations that involve oxygen would be natural extensions of this process, with O_2 either playing the role of A or reacting with A^- to give O_2^-. With an H^+ ion the O_2^- could form an HO_2 radical.

Many reactions of the type shown (5-13) can be demonstrated in the absence of O_2. With the dye eosin as sensitizer and phenol as electron donor, L. Grossweiner has shown that excitation of the eosin produces an oxidized phenol radical and reduced eosin:

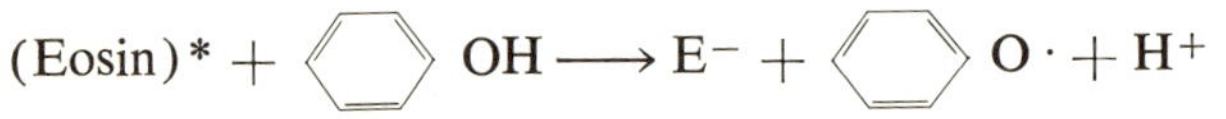

(5-14)

This can be followed by transfer of an electron from the eosin to any of a variety of acceptors including O_2. But if eosin is illuminated in a solution with ferricyanide, electrons are transferred from the dye to the ferricyanide, leaving oxidized eosin which can subsequently oxidize an electron donor.

If we examine reactions of Chl in vitro, we find that these can give either oxidized or reduced Chl as an intermediate, depending on the nature of the other reactants. In the presence of O_2 and ascorbic acid, the first reaction is usually one in which Chl is oxidized and O_2 is reduced. Ascorbic acid then reduces the oxidized Chl. But if O_2 is replaced by methyl viologen as electron acceptor, the first reaction is usually between Chl and ascorbic acid, giving reduced Chl, which can in turn reduce the methyl viologen.

Many other reactions of these kinds have been observed. In most of the cases that have been studied well, the singlet excited state of the sensitizer is translated into a triplet state before the electron transfer happens. There are several lines of evidence for this: First, the reaction is usually dependent on collisions between the reactants, and by comparing the quantum efficiency with the frequency of collisions it can be concluded that the excited state must be relatively long-lived, perhaps a millisecond or longer.[1] An excited state lasting this long is most likely a triplet state.

Further evidence comes from transient absorption spectra that are characteristic of sensitizers in their triplet states. Finally there is the fact that triplet excited states are quenched strongly by O_2, by carotenoids (provided they have nine or more conjugated double bonds), and by paramagnetic ions such as Fe^{++} or Mn^{++} (but not by diamagnetic ones such as Mg^{++} or Zn^{++}). These agents usually interfere with sensitized electron transfer to the extent that they can quench the triplet excited state of the sensitizer.

Reactions of the kind $S^* + O_2 \longrightarrow S^+ + O_2^-$ have been documented often as examples of photochemical electron transfer. However, a different kind of reaction will supervene if a suitable substrate is present. The reaction involves a sensitizing pigment, oxygen, and the substrate. Leaving out the photochemical details for the moment, the reaction between oxygen and substrate is typified by the following examples:

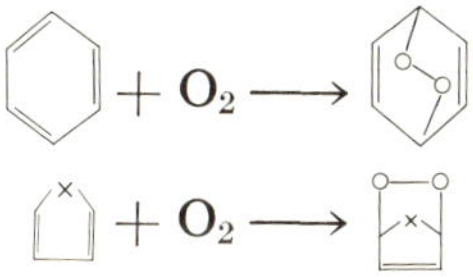

(5-15)

Among molecules of biological interest, favorable substrates for these oxygenations include histidine and the aromatic amino acids, and bases of nucleic acids. These reactions attain quantum efficiencies approaching one O_2 per quantum, far greater than the efficiencies for photochemical transfer of electrons to O_2. Therefore the oxygenation predominates if a suitable substrate is present.

Denoting the substrate as R, the sensitizer as S, and excitation by an asterisk, the following mechanisms have been entertained for the broad outlines of the photochemistry:

[1] This kind of reasoning was developed more fully in Sec. 3-10 and Prob. 15 in Volume 1.

$$S^* + O_2 \longrightarrow (S \cdot O_2)^* \qquad R + (S \cdot O_2)^* \longrightarrow S + RO_2 \qquad (5\text{-}16)$$
$$S^* + O_2 \longrightarrow SO_2 \qquad R + SO_2 \longrightarrow S + RO_2 \qquad (5\text{-}17)$$
$$S^* + O_2 \longrightarrow S + O_2^* \qquad R + O_2^* \longrightarrow RO_2 \qquad (5\text{-}18)$$

The third of these alternatives, posing excited oxygen as an intermediate, was supported strongly by ingenious experiments made in the 1930s by H. Kautsky. He adsorbed the sensitizer on one set of silica particles and the substrate on another. Then he mixed the particles, illuminated them, and observed a high yield of oxygenated substrate. The sensitizer was not changed by the reaction. This could only mean that the reaction was carried from sensitizer to substrate by the long-range migration of a diffusible substance: not $(S \cdot O_2)^*$, not SO_2, but presumably O_2^*.

Kautsky's view of photochemical oxygenation was unpopular for many years, probably because of ignorance about the excited states of oxygen. We shall see in the next section that his interpretation appears to be correct.

5-10. Photochemical Oxygenation: The Nature and Importance of Excited Oxygen

For most molecules the ground state is a singlet (total electron spin $S = 0$). Because the spin is not likely to change during a transition, the excited states formed most commonly are also singlet. If a triplet state ($S = 1$) is once formed, it has a relatively long lifetime, milliseconds to seconds, because the probability (per unit time) of returning to the singlet ground state is low.

With O_2 and other paramagnetic molecules this situation is reversed: the ground state is a triplet, and among excited states the singlets are the metastable and therefore long-lived ones. Singlet excited states of O_2 have mean lifetimes of several seconds and will react well with appropriate substrates to form organic peroxides.

Evidence for excited oxygen comes from characteristic absorption and emission of light. In 1927 L. Mallet observed a flash of red light[1] when he mixed hydrogen peroxide with hypochlorite (OCl^-, as found in laundry bleach). This chemiluminescence was noticed again in 1960 by H. H. Seliger, who identified the wavelength of the red emission as 633 nm. Subsequently M. Kasha suggested that the emission comes from a singlet excited state of oxygen. Confirming another observation by Mallet, he found that if a dye such as eosin is present in the reaction mixture, the characteristic fluorescence of the dye is seen. This could be attributed to a transfer of energy from excited oxygen to the dye (transfer in the reverse direction, from the dye triplet state, is important in photochemical oxygenation and will be discussed later). The several seconds' lifetime of the excited state of oxygen could be shown strikingly by bubbling the gases from the reaction mixture into a second vessel containing a dye, and observing fluorescence of the dye.

Extension of this work showed that the reaction of peroxide with hypochlorite generates light at 480, 762, and 1,270 nm as well as at 633 nm. Emission at these same wavelengths was observed by R. J. Browne and E. A. Ogryzlo when they passed an electric discharge through pure O_2. This proved that the chemiluminescent emissions represent states of oxygen. It was well known that liquid oxygen is blue, which means that it absorbs red light, and upon closer examination oxygen proved to have absorption bands matching all the foregoing emission bands: 480, 633, 762, and 1,270 nm, and also some in the UV. These bands represent states, then, that can be reached either by the direct absorption of light (see Volume 1, Chap. 2) or by a chemical reaction or electric discharge.

The bands at 1,270 and 762 nm are predicted in a routine application of quantum theory to O_2 molecules. The ground state of O_2 is called $^3\Sigma$; the Σ is a symbol signifying a state with axial symmetry, and the superscript 3 means triplet. Two singlet excited

[1] This flash can be observed with dark-adapted eyes when household peroxide (3% H_2O_2) and bleach solution are mixed. Be prepared for a violent effervescence and release of chlorine; protect your eyes—do not hold the mixture above eye level.

states whose energies correspond to 1,270 and 762 nm are called $^1\Delta$ and $^1\Sigma$ respectively (Δ represents another symmetry of the electronic wave functions, distinct from Σ). The energies of these states are sketched in Fig. 5-7. The wave number k (reciprocal of the wavelength in centimeters) is proportional to energy above the ground state.

The higher-energy states, shown by the emission and absorption at 633 and 480 nm, defied understanding until several in-

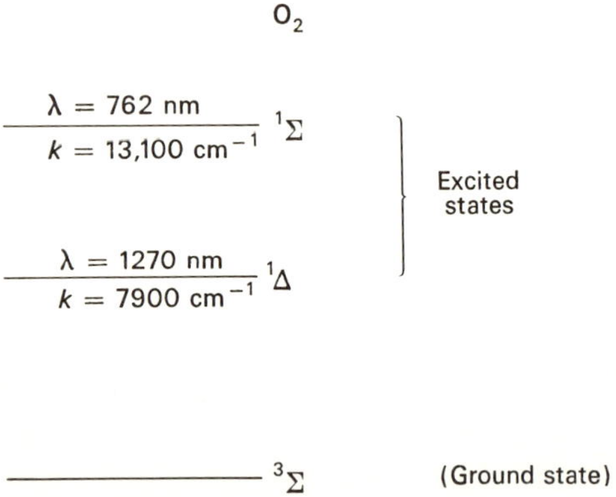

Fig. 5-7. Ground and excited state of O_2 *(see the text). The wave number k,* cm^{-1}*, is the reciprocal of the wavelength and measures energy above the ground state: 8,100* cm^{-1} *is equivalent to 1,240 nm, 1 electron volt, or 23 kcal/mole.*

vestigators including Browne and Ogryzlo suggested that they are formed by the interaction of *two* excited O_2 molecules. Thus the 633-nm state has the energy of two $^1\Delta$ molecules:

$$2 \times O_2^* \ (^1\Delta, \lambda = 1{,}270 \text{ nm}, k = 7900 \text{ cm}^{-1}) \longrightarrow O_4^* \ (\lambda = 633 \text{ nm}, k = 15{,}800 \text{ cm}^{-1}) \qquad (5\text{-}19)$$

and the energy of the 480 nm (21,000 cm^{-1}) state is just the sum of the energies of $^1\Delta$ and $^1\Sigma$:

$$O_2^* \, (^1\Delta, k = 7900 \text{ cm}^{-1}) + O_2^* \, (^1\Sigma, k = 13{,}100 \text{ cm}^{-1}) \longrightarrow O_4^* \, (\text{k} = 21{,}000 \text{ cm}^{-1}) \qquad (5\text{-}20)$$

Such bimolecular reactions can happen because of the long lifetimes of the excited (O_2^*) molecules. This explains the emissions at 480 and 633 nm, and the sensitized fluorescence (mediated by excited oxygen) of dyes in this wavelength range. The absorptions of light at 480 and 633 nm in liquid oxygen must be visualized as dependent on concerted four-atom events giving O_4^*.

Excitation can pass from an excited sensitizer to oxygen,

$$S^* + O_2 \longrightarrow S + O_2^*$$

If the sensitizer has first entered a triplet state, the subsequent formation of singlet O_2^* can be written

$$S^{*(T)} + O_2 \text{ (ground, triplet)} \longrightarrow S \text{ (ground, singlet)} + O_2^{*(S)} \qquad (5\text{-}21)$$

As pointed out by D. R. Kearns and A. U. Khan (see the Bibliography) this reaction involves an exchange of electrons between S and O_2, but no change in the spin orientations of the electrons. These remarks are illustrated in Fig. 5-8, where we keep track of electrons and their spin orientations throughout the process. If the electron transfer from S^* to O_2 should happen before the reverse one, the pair S^+ and O_2^- would exist as transient intermediates. This shows a relationship between Reaction (5-21) and the photochemical electron transfer discussed earlier.

Reaction (5-21) is a collision-dependent process with high probability per collision between S^* and O_2. Kearns has observed directly the formation of $O_2^{*(S)}$ from triplet excited naphthalene or benzene, with a quantum efficiency about 50 percent. This kind of reaction may account partly for the efficiency with which oxygen quenches triplet states of dyes.

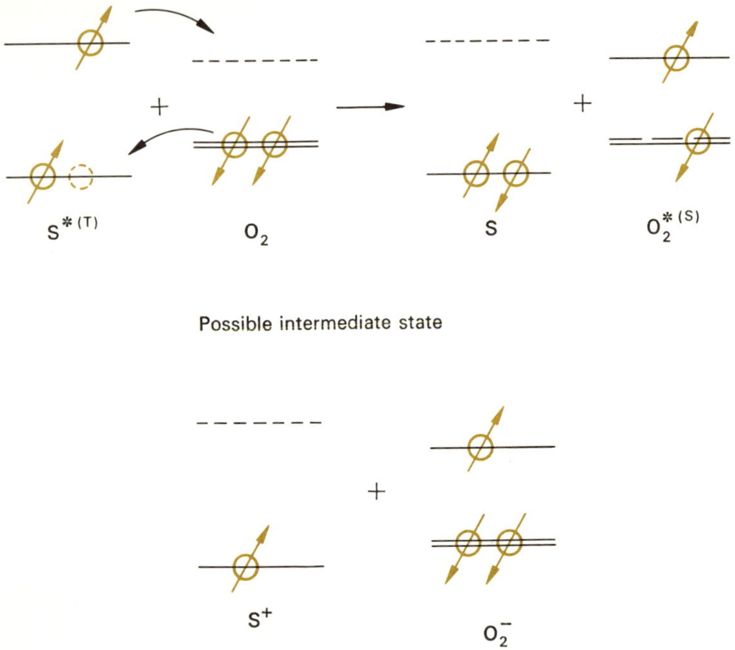

Fig. 5-8. Triplet excited sensitizer can exchange electrons with triplet (ground state) O_2 *to form singlet (ground state) sensitizer and singlet (excited)* O_2. *This reaction does not require a change of electron spin. Oxidized sensitizer and* O_2^- *could conceivably exist as transitory intermediates in the reaction.*

In most photochemical reaction mixtures O_2 merely quenches the triplet excitation of the sensitizer and thus inhibits other photochemical events. But with certain favorable substrates, such as those in Reaction (5-15), oxygenation occurs with high quantum efficiency if the concentration of O_2 is greater than about $10^{-5}M$ (in water saturated with air at room temperature, the concentration of O_2 is about $2 \times 10^{-4}M$). This reaction is surely of great importance in photooxidative damage to living systems, since components of both proteins and nucleic acids (especially histidine and purines) are good substrates and the concentration of O_2 is high enough in most cells and tissues of aerobic creatures.

The importance of singlet oxygen in sensitized photooxygenations is shown by two major sets of experiments, in which reactions of singlet oxygen prepared chemically or by electric discharge have been compared with dye-sensitized reactions. C. S. Foote prepared singlet oxygen by the reaction of peroxide with hypochlorite and allowed the excited oxygen to react with a variety of unsaturated hydrocarbons. For comparison he observed dye-sensitized photooxidations of the same hydrocarbons. Foote studied about 30 hydrocarbons, each giving more than one reaction product. In every case the distribution (relative yields) of the products was the same for the photochemical reaction as for the reaction with excited oxygen produced chemically. The same thing was shown by E. J. Corey and K. D. Bayes, using excited oxygen prepared by an electric discharge.

The protective action of carotenoid pigments has been discussed in terms of the ability of carotenoids to quench triplet states of dyes. Foote has pointed out that carotenoids can quench the excited states of oxygen, without themselves being changed chemically. Those carotenoids (with nine or more conjugated double bonds) that are good quenchers of triplet excitation energy in dyes are also good quenchers of the singlet energy in excited oxygen. It may be that the carotenoids protect living things against photooxidative damage by more than one mechanism.

The high specificity of organic oxygenations can be understood in terms of physicochemical principles that are rooted in the conservation of momentum, charge, mass, and energy. These principles dictate the allowedness or forbiddenness of transition between different states of an atom or molecule, and they govern chemical reactions as well. In the special context of the reactions between oxygen and organic molecules, these principles reduce to three major requirements for a high probability of reaction: (1) The reaction must be exothermic; this usually means that O_2 must enter the reaction in an excited state. (2) The total electron spin of the products must equal that of the reactants. (3) The symmetry of the system must not change during the reaction.

Symmetry is a property of the wave functions [1] of all the elements of the system. It is defined by the sign of the total wave function, relative to the center of mass of the entire system consisting of reactants or products. Put the center of mass at the origin of a set of coordinate axes; in this framework every point has coordinates x, y, and z. Now a wave function is symmetric, with respect to the center of mass, if its value at any point (x, y, z) is the same as at the point $(-x, -y, -z)$. It is antisymmetric if the wave function at $(-x, -y, -z)$ has the same magnitude, but opposite sign, as at (x, y, z). These conditions of symmetry and antisymmetry are called even and odd parity, respectively. An actual wave function can be made up by superimposing perfectly symmetric and antisymmetric functions. The requirement that the parity must not change in a reaction is basically a requirement that momentum be conserved.

The consequences of requiring exothermicity, conservation of spin, and conservation of parity or symmetry are discussed for organic oxygenations in an article by Kearns and Khan; see the Bibliography. Without going into details, let us consider one simple interplay between conservation of spin and the energy change of a reaction between O_2 and an organic molecule R. Write the reaction $O_2 + R \longrightarrow RO_2$, and consider first that the substrate R is in its ground state (a singlet state), and the O_2 is in its ground (triplet) state. Then if electron spin is to be conserved, the product must be formed in the triplet state:

$$O_2 \text{ (triplet, ground)} + R \text{ (singlet, ground)} \longrightarrow RO_2 \text{ (triplet)} \qquad (5\text{-}22)$$

But the triplet state of RO_2 is nearly always an excited state, so this reaction would usually be strongly endothermic. On the other hand

[1] Recall from Sec. 2-4, Volume 1, that the wave function $\psi(x, y, z)$ of a particle is a map of the particle's position. The quantity $\psi^2\, dV$ gives the probability of finding the particle in a small volume dV located at coordinates (x, y, z).

if the O_2 enters the reaction in an excited (singlet) state, the product RO_2 can be formed in a singlet state, and if this is the ground state of the product, the reaction may be highly exothermic.

The conservation of symmetry often imposes an added restriction: the lowest singlet state of O_2, the $^1\Delta$ state, can react with the substrate to give the product in its singlet ground state, but the higher excited state of O_2, $^1\Sigma$, must lead to product in a singlet *excited* state. This again would be an endothermic process. This explains why the less energetic of the two singlet states of O_2 is usually the more reactive.

These considerations have given us much insight into the mechanisms by which light and oxygen cooperate to cause damage to living things. The same principles applied to similar chemical reactions are part of a recently improved understanding of bioluminescence.

BIBLIOGRAPHY

1. Watson, J. D.: "Molecular Biology of the Gene," W. A. Benjamin, Inc., New York, 1965. A clear and well-balanced account of the biochemistry of nucleic acids and proteins and of genetics at the molecular level.
2. Jagger, J.: "Introduction to Research in Ultraviolet Photobiology," Prentice-Hall, Inc., Englewood Cliffs, N.J., 1967. An exceptionally fine introduction to the methods and current state of knowledge of this subject.
3. Oparin, A. I.: "Life: Its Nature, Origin and Development," Academic Press, Inc., New York, 1964. The classic account of modern speculations on the origin of life on the Earth.
4. Berkner, L. V., and L. C. Marshall: The History of Oxygenic Concentration in the Earth's Atmosphere, *Discussions Faraday Soc.,* vol. 37, pp. 122–141, 1964.
5. Hanawalt, P. C.: Radiation Damage and Repair *in vivo,* in C. P. Swanson (ed.), "An Introduction to Photobiology," p. 53, Prentice-Hall, Inc., Englewood Cliffs, N.J., 1969.
6. Wiskemann, A.: Effects of Ultraviolet Light on the Skin, in

C. P. Swanson (ed.), "An Introduction to Photobiology," p. 81, Prentice-Hall, Inc., Englewood Cliffs, N.J., 1969.

7. Blum, H. F.: "Photodynamic Action and Diseases Caused by Light," Reinhold Publishing Corporation, New York, 1941. A classic compendium of this subject.
8. Macalpine, I., and R. Hunter: Porphyria and King George the Third, *Sci. Am.*, vol. 221, p. 38, July, 1969. An interesting delineation of this disease and its effects on the royal families of Europe.
9. Kearns, D. R., and A. U. Khan: Sensitized Photooxygenated Reactions and the Role of Singlet Oxygen, *Photochem. Photobiol.*, vol. 10, p. 193, 1969.

6.
The Luminescence of Fireflies and Other Living Things

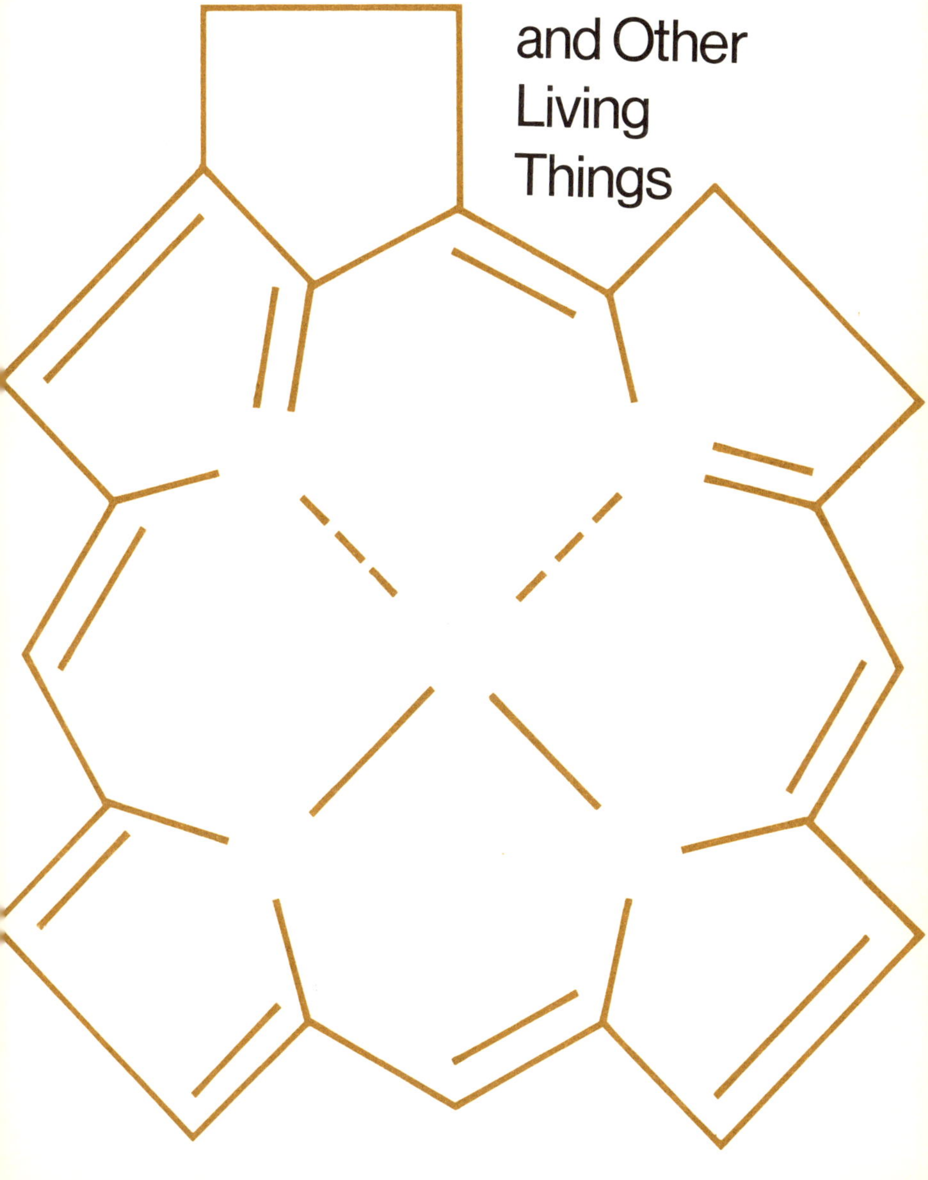

The property of luminescence is distributed seemingly at random among types of animals and microbes, although it is far more common in marine organisms than in terrestrial or freshwater varieties. Luminescence has been observed and studied especially in bacteria, protozoa, fungi, jellyfish, clams, worms, and arthropods. Our most detailed knowledge pertains to the mechanism of firefly luminescence. Our partial understanding of other systems reveals considerable chemical diversity. In contrast to the rule of biochemical unity, arising from a shared history of biochemical evolution for most life processes, the development of luminescent systems seems to have happened independently and in many different ways in different forms of life.

The apparent biological advantages of luminescence are also diverse, and sometimes obscure. The usefulness of a special sequence of light flashes is obvious in connection with mating habits and recognition of species among fireflies. On the other hand the value of luminescence to a mold is not clear at all.

We shall survey the chemical outlines of bioluminescence in a variety of cases, treating the firefly system more thoroughly as befits our more detailed knowledge. We shall finish with a brief consideration of the biological significance of luminescence.

DIVERSE CHEMICAL PATTERNS OF BIOLUMINESCENCE

Most bioluminescent processes involve the enzyme-catalyzed reaction of a substrate with O_2. To follow the terminology coined by R. DuBois in the 1880s, enzymes involved in bioluminescent reactions are called *luciferases,* and the substrates, although quite diverse in different creatures, carry the generic title *luciferins.*

6-1. Luminous Crustaceans, Protozoa, Fungi, and Sea Pansies

A tiny crustacean found in Japanese waters, *Cypridina,* squirts luciferin and luciferase into the sea, where they mix and react with O_2 to give a flash of light. The luciferase is a protein of molecular weight about 50,000. It lacks any obvious prosthetic group, metal, or other special component that could serve as a light-emitting chromophore. The luciferin has a known structure that was confirmed through chemical synthesis by Y. Kishi and collaborators; it is shown in Fig. 6-1. The reaction of the luciferin with O_2 consumes one O_2 per molecule of luciferin in a reaction that is probably like that of firefly luciferin (see later).

The emission spectrum of *Cypridina* luminescence has a peak at 460 nm. The emitting species is not known, but it ought to be some form of the luciferin resulting from its interactions with the luciferase and O_2. The reaction product, oxyluciferin, has a fluorescence maximum at 480 nm, which could conceivably be shifted to 460 nm in association with the enzyme (although the binding of

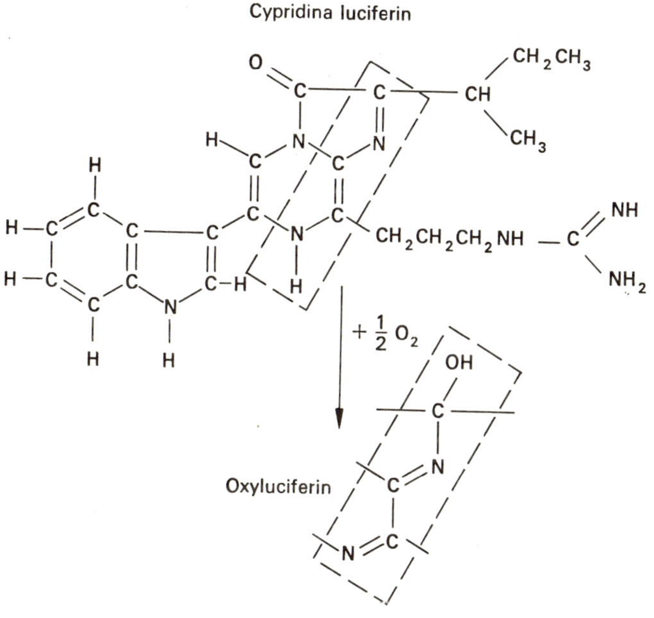

Fig. 6-1. Structure of the substrate luciferin in the bioluminescent reaction of the marine crustacean Cypridina. *Luciferin is a generic term; in another organism it can refer to a different molecule (compare Fig. 6-3).*

chromophores to proteins usually causes a shift toward greater wavelengths). Details of the reaction mechanism remain unknown. Since the product is dissipated in the sea, the question of biochemical recycling of the luciferin does not arise.

J. W. Hastings and collaborators have studied an interesting case of luminescence in *Gonyaulax,* a marine protozoan (see Sec. 4-9, describing circadian rhythms in this organism). A soluble luciferin-luciferase system can be isolated from this creature, but within the cells the system appears to be bound to the surfaces of solid inclusions called *scintillons.* The scintillons can be isolated from broken cells by density gradient centrifugation. They can be "discharged," or made to emit light, simply by lowering the surrounding pH from 8 to 5.7. This luminescence, like the others described in this section, requires O_2. The time course and spectrum of the emission induced in a suspension of scintillons mirror those of the emission by the intact cells.

The scintillons of *Gonyaulax* may serve to bring the luciferin and luciferase together at high concentrations and to regulate the flashing by some sort of control that involves H^+ ions.

In some cases of luminescence the luciferin engages in a recognizable preparatory step before the light reaction. Fungal luciferin is reduced by NADH to become active:

$$L + NADH + H^+ \longrightarrow LH_2 + NAD^+ \qquad (6\text{-}1)$$

The reduced luciferin then reacts with O_2 in the presence of luciferase to give light. The fungal luminescence, studied extensively by R. L. Airth, is continuous, and its intensity is usually limited by the availability of substrates that can reduce NAD^+ to NADH.

The sea pansy *Renilla* is an animal whose luminescence has been investigated by M. J. Cormier and J. R. Totter. The luciferin occurs in an inactive form bearing a sulfate group. Activation is an enzyme-catalyzed, Ca^{++}-dependent transfer of the sulfate to 3′, 5′-diphosphoadenosine (DPA, which is the same as "ordinary" adenosine-5′-monophosphate, AMP, with an extra phosphate attached to the ribose part). Once prepared, the luciferin reacts with O_2 in a luminescent reaction catalyzed by a second enzyme, the luciferase. In this system the preparatory step might provide a mechanism for the timed control of luminescence.

We shall see that in firefly luminescence a single enzyme, the luciferase, catalyses both a preparatory step (involving ATP) and the luminescent reaction.

6-2. Luminous Bacteria

The luminescent bacteria have received considerable study in recent years by J. W. Hastings, Q. H. Gibson, and associates. The light-yielding reaction is the culmination of a sequence of reactions involving NADH (refer to Fig. 1-3), flavine mononucleotide (FMN), a luciferase, O_2, and a long-chain aliphatic aldehyde. FMN is an important coenzyme in biological oxidations; its structure (together with the reduced form $FMNH_2$) is shown in Fig. 6-2.

The first recognized step in bacterial luminescence is the reduction of FMN by NADH, mediated by the enzyme FMN reductase:

$$FMN + NADH + H^+ \xrightarrow{\text{enz}} FMNH_2 + NAD^+ \qquad (6\text{-}2)$$

The $FMNH_2$, in association with luciferase, then reacts with O_2 to form an intermediate of several seconds' lifetime. The intermediate decays to an unknown set of products in a reaction that gives light. The nature of this intermediate "$FMNH_2$-enzyme-oxygen" compound is not known. One possibility is that the $FMNH_2$ reduces one site on the enzyme, and the O_2 oxidizes another site. Interaction of the oxidizing and reducing parts of the enzyme could then generate an excited state (of what chromophore?), in a process loosely analogous to the delayed fluorescence of Chl in plants.

The need for $FMNH_2$ can be bypassed by illuminating the enzyme protein (luciferase) with 280-nm light. This produces a

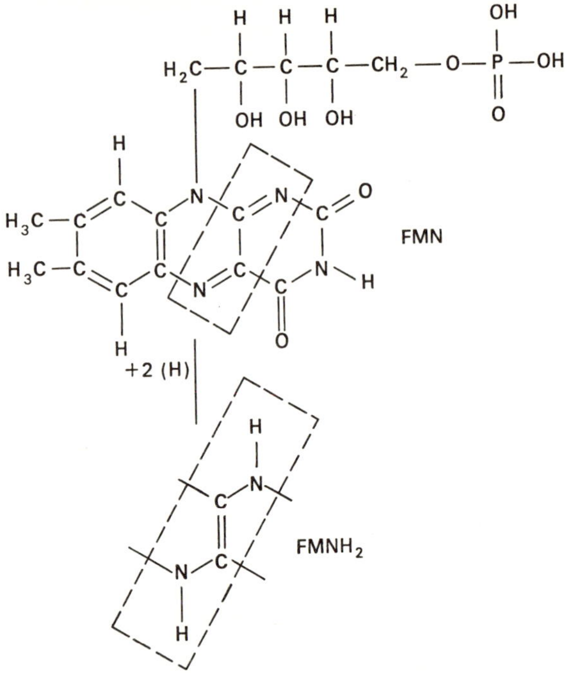

Fig. 6-2. Flavine mononucleotide (FMN) and its reduced form $FMNH_2$: a compound of isoalloxazine, ribitol (not ribose), and phosphate. This may serve as the luciferin in bacterial luminescence.

state which can interact with O_2 and give apparently the same intermediate, precursor of the luminescent reaction, that is formed ordinarily by $FMNH_2$, enzyme, and O_2.

A chemical peculiarity of the bacterial luminescent reaction is its dependence on the presence of a long-chain (e.g., 8-carbon to 16-carbon) aldehyde. When the luminescence is studied in vitro, using purified luciferase together with $FMNH_2$ and O_2, little or no light is emitted unless aldehyde is added. Without aldehyde the intermediate decays by a "dark" route, as if there were a quenching process that is prevented by aldehyde:

$$\left.\begin{array}{l} FMNH_2 \\ \text{Luciferase} \\ O_2 \end{array}\right\} \rightarrow \text{“intermediate”} \left\{\begin{array}{l} \xrightarrow{\text{(aldehyde)}} \text{products} + \text{light} \\ \\ \longrightarrow \text{products} \end{array}\right. \qquad (6\text{-}3)$$

The entity whose excited state is responsible for emission in luminous bacteria has not been identified. One might suppose it to be some state of FMN, or FMN combined with aldehyde, or an oxidation product thereof, perhaps modified by association with the luciferase. The bioluminescence is maximal at 490 nm. FMN has a fluorescence maximum at 530 nm, but this might be shifted by the binding to enzyme or aldehyde. On substitution of an alternative substrate, *iso*-FMN, the peak of bioluminescence is shifted from 490 to 470 nm, whereas the fluorescence maximum of *iso*-FMN is at 550 nm.

Clearly there is much to be learned about this interesting system.

6-3. Bioluminescence without Oxygen: Acorn Worms and Jellyfish

Plants contain peroxidases, enzymes that mediate the oxidations of various substrates by H_2O_2. E. Newton Harvey, a pioneer in the study of luminescence during this century, recognized decades ago that with certain substrates the peroxidatic reaction is accompanied by the emission of light. These photogenic substrates include pyrogallol and luminol (3-aminophthalhydrazide). This seemed to be a trivial case of bioluminescence in plants, unrelated to animal bioluminescence, until a similar process was recognized in a marine animal by L. S. Dure and M. J. Cormier. The animal is a hemichordate called the *acorn worm.* It gives luminescence under the stimulus of H_2O_2, and to some extent its components (luciferin and luciferase) can be interchanged with their counterparts in the plant peroxidase system. Thus if a partial purification is made of the luciferin and luciferase from acorn worms, the luciferin fraction will give light when mixed with H_2O_2 in the presence of peroxidase from horseradish, and conversely certain substrates of horseradish peroxidase will react with the worm luciferase and H_2O_2 to give light.

In a way the simplest (and one of the most obscure) luminescent systems is that of some jellyfish: a protein that emits light when stimulated by the addition of Ca^{++} ions. In this system, described by O. Shimomura, F. H. Johnson, and Y. Saiga, the chemical requirements for emission appear to be built up on the protein, and Ca^{++} then causes the components to react and emit light. This discharge has a half-life of about a second. The active protein has a firmly bound prosthetic group spectrally similar to NAD. The Ca^{++}-triggered reaction results in the appearance, in the protein, of an absorption band at 340 nm, typical of NADH. Along with this new absorption band, the protein acquires a fluorescence band centered at 460 nm, also typical of NADH. The spectrum of the Ca^{++}-induced luminescence is also like this fluorescence, showing a maximum at 460 nm. Therefore it is likely that the luminescent reaction involves the reduction of protein-bound NAD^+ to NADH, with the reduced product appearing in its excited state and emitting light as it returns to the ground state. Of course, these remarks do not explain the mechanism by which Ca^{++} causes such a reaction on the protein.

6-4. Firefly Luminescence

The glow of the firefly comes from an organ in the abdomen. The timing of the flash sequence is under nervous control; a decapitated firefly glows steadily for some time.

In different species of firefly, emission varies from green to orange, with maxima ranging from about 550 to 595 nm. In one exceptional species the individuals from different locales show a variation from 576 to 594 nm. These variations have been associated conclusively with differences in the luciferase rather than the luciferin, as shown by cross-reacting the enzymes and substrates from different species. The changes in wavelength are ascribed with customary (and necessary) vagueness to different effects of the enzyme on the shape and electrical properties (electron distribution; ionization) of the enzyme-bound substrate.

The mechanism of firefly luminescence has been a main topic of study at the McCollum-Pratt Institute of Johns Hopkins University. Our knowledge of this subject has come largely from the efforts of the institute's director, W. D. McElroy, and many of his associates, in recent years especially T. A. Hopkins, F. McCapra, H. H. Seliger, and E. H. White.

The total synthesis of firefly luciferin has confirmed the structure shown in Fig. 6-3. The luciferase has been crystallized and characterized extensively; it has no outstanding metals or prosthetic groups that could act as chromophores for the luminescence.

The luminescent reaction requires a preparatory step that bears a formal resemblance to the activation of amino acids in preparation for protein synthesis. An enzyme (in this case the

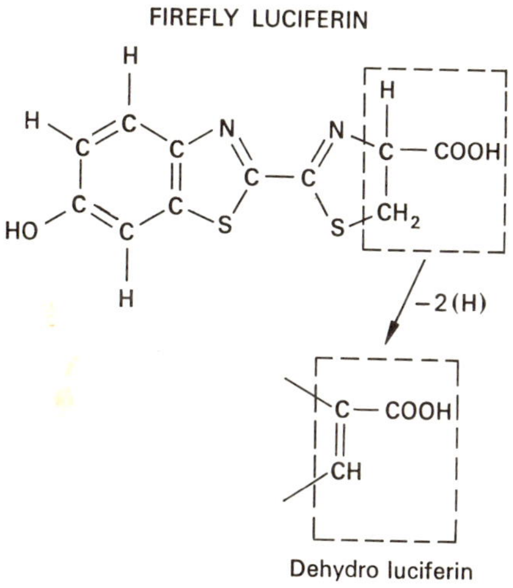

Fig. 6-3. Structures of firefly luciferin and a derivative, dehydroluciferin, that is not part of the luminescent reaction (compare Fig. 6-4). The part of the molecule enclosed by dashed lines undergoes the reactions shown in Fig. 6-4.

luciferase) catalyzes a reaction between luciferin and ATP, forming a luciferin-AMP compound and releasing inorganic pyrophosphate: [1]

$$LH_2 + E + ATP \xrightarrow{(Mg^{++})} E \cdot LH_2 \cdot AMP + PP_i \qquad (6\text{-}4)$$

Here E stands for the enzyme and PP_i for inorganic pyrophosphate. The luciferin has been abbreviated LH_2 for historical reasons. The abbreviation L is reserved for a dehydro derivative, with two H atoms less. This derivative, also shown in Fig. 6-3, was once thought to be a product of the bioluminescent reaction. We now know that it is not.

The preparatory step shown in Reaction (6-4) is catalyzed by the same enzyme (luciferase) that mediates the luminescent reaction. The requirement for ATP is specific; other common nucleoside triphosphates are inactive. The firefly luminescent reaction is therefore a highly sensitive and specific test for the presence of ATP in a sample. The sample can be mixed with an aqueous extract of firefly tails, and a flash of light resulting from about 10^{-10} mole of ATP can be measured without difficulty by means of a photomultiplier. It is curious that this assay, based on luminescence in fireflies, led accidentally to the discovery (Sec. 1-10) of Chl luminescence or delayed fluorescence in plants.

Although the attachment of AMP to a substrate at the expense of ATP is generally described as an activation of the substrate, the term is a little misleading in this case, because the energy introduced by this reaction (a few kilocalories per mole) falls far short of the amount (about 50 kilocalories per mole) released as

[1] Recall that pyrophosphoric acid is a conjunction of two molecules of ortho-phosphoric acid:

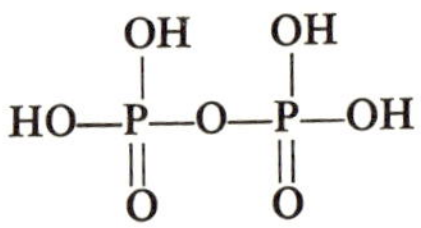

yellow-green light. The energy gained from ATP in the preparatory step is not at all a major source of the light energy.

Once the structure of luciferin and the details of the preparatory reaction had become known, the way was paved to identify the probable intermediates and products of the O_2-dependent luminescent reaction. This work was aided by the fact that firefly luciferin and closely related molecules can engage in chemiluminescent reactions, with O_2 but without enzyme, in anhydrous media (dissolved in dimethyl sulfoxide). Comparison with these reactions helped to identify the products of the bioluminescent reaction and to construct a highly plausible reaction sequence.

One possible mechanism is outlined in Fig. 6-4. The first step is the preparatory one [Reaction (6-4)]. The next step is analogous to the oxygenation reactions described in the last chapter, except that O_2 enters the reaction in its ground state. The luciferin-AMP compound incorporates O_2 as an endoperoxide while discharging the AMP. After this the molecule splits, releasing CO_2 and leaving the main reaction product, decarboxyketoluciferin, in an excited state. A quantum of light is emitted as this molecule subsides to the ground state. The likelihood of the successive steps, including the formation of the product in its excited state, could probably be justified by appealing to the conservation of spin and symmetry as described in the last chapter. The details of such a justification have not been traced fully and would be far beyond the scope of this book. Suffice it to say that the reactants contain enough potential energy to account for the quantum of light emitted. Also it is worth noting that this could be a relatively benign way for an organism to deal with the oxygenating tendency of O_2.

The emitting molecule, decarboxyketoluciferin, can also exist in an enol isomeric modification, and both the keto and enol forms can be ionized, as shown in Fig. 6-5 (one possibility, the enol monoanion, is left out). The enol dianion in an excited singlet state has been identified presumptively as the emitting species in the yellow-green luminescence of the living firefly, because the fluorescence of this dianion has a spectrum that matches the spectra of bioluminescence and chemiluminescence. In several derivatives or

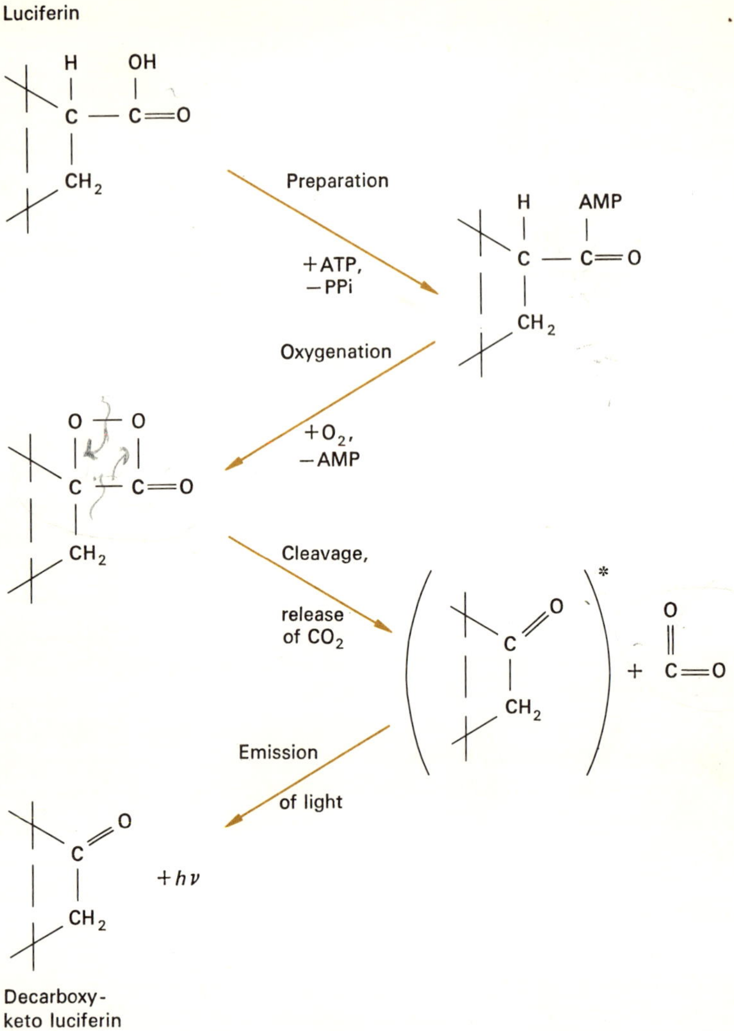

Fig. 6-4. An outline of a possible mechanism of firefly luminescence, omitting the role of the enzyme luciferase (see the text). The substrate luciferin is first prepared by attachment of adenosine monophosphate, AMP; see Reaction (6-4). Next it is oxygenated with release of AMP and then cleaved with release of CO_2. *The main reaction product, decarboxyketoluciferin, is formed in a singlet excited state and emits a quantum of light as it goes to its electronic ground state. Most of the luciferin molecule has been left out of these drawings; compare Fig. 6-3.*

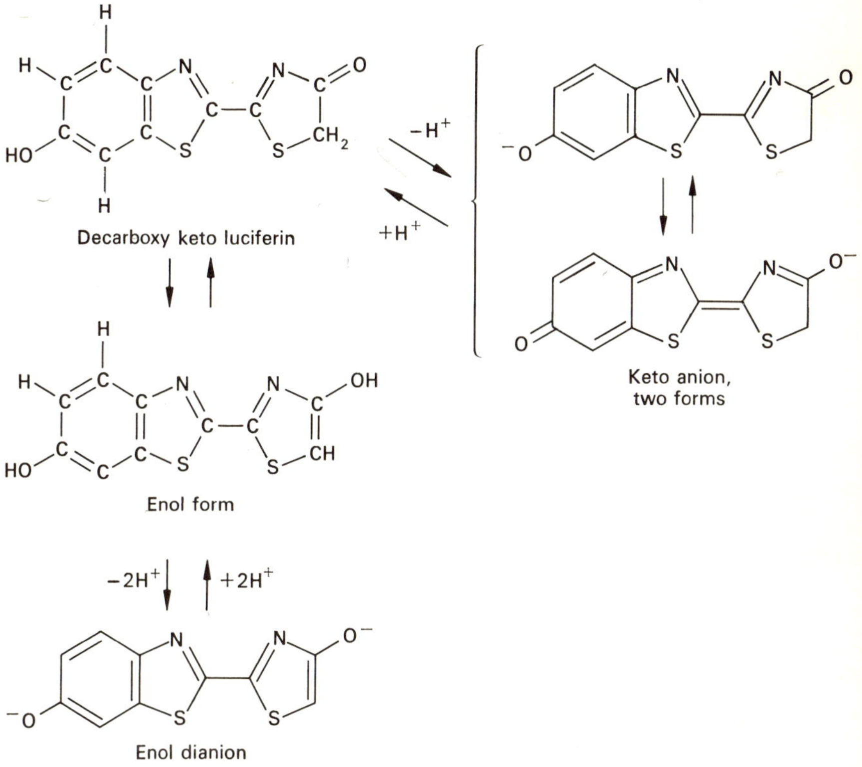

Fig. 6-5. The product of the firefly luminescent reaction, decarboxyketo-luciferin, can exist in keto and enol isomeric forms, and these can be ionized as shown. The enol dianion emits the yellow-green light typical of the living firefly; the keto anion (resonating between the two structures shown) emits red light.

analogs of this molecule the fluorescence band of the enol dianion is shifted to new wavelengths, and the maxima of chemiluminescence and bioluminescence are shifted correspondingly. This helps to confirm the enol dianion as the emitter of bioluminescence, in the face of the fact that the maximum wavelength of bioluminescence is shifted by some influence of the luciferase.

Not only is the emission spectrum of the enol dianion of decarboxyketoluciferase shifted a little by the enzyme; the equilibrium between this and other forms of the molecule as shown in Fig. 6-5 is affected. For a particular much-studied species of firefly, *Photinus pyralis,* the emission in vivo (identified with the excited state of the enol dianion) peaks at 562 nm. If the bioluminescent reaction of this species is studied in vitro, with purified luciferin and luciferase, it is found that the yellow-green (562 nm) emission typical of the living firefly occurs if the pH of the reaction mixture is slightly alkaline (above about 7.5). To the acid side of neutrality a red emission, with a peak at 614 nm, replaces the yellow-green emission. This is due to the keto anion form of the decarboxy-ketoluciferase (Fig. 6-5). The state of this molecule is governed by the luciferase. First, the equilibrium between keto and enol forms can probably be influenced by chemical interactions with groups such as histidine in the enzyme. Second, the buffering action of groups such as -SH and $-NH_2$ can affect the ionization. These chemical details are discussed by McElroy, Seliger, and White (see the Bibliography).

Although dehydroluciferin (see Fig. 6-3) is not in the bio-luminescent reaction sequence, the compound of this molecule with AMP binds firmly to the active site of luciferase and thus makes the enzyme unavailable for the bioluminescent reaction. The L·AMP can be removed from the enzyme, both by pyrophosphate

$$E \cdot L \cdot AMP + PP_i \longrightarrow E + L + ATP \qquad (6\text{-}5)$$

in the reversal of an "activating" step and by ATP, which induces the enzyme to act in a hydrolytic [1] way:

[1] In Reaction (6-6), and also in the first two steps of Fig. 6-4, the reader may be troubled by an apparent lack of chemical balance for —H, —OH, and H_2O. Following custom, we have omitted certain details concerning these entities. Reaction (6-6) should strictly be written

$$E \cdot L \cdot AMP + H_2O \longrightarrow E + L\text{-}OH + H\text{-}AMP$$

but the L-OH as written here is ordinarily called L, and the H-AMP is written AMP.

$$E \cdot L \cdot AMP + H_2O \xrightarrow{(ATP)} E + L + AMP \qquad (6\text{-}6)$$

These reactions, by governing the availability of the enzyme for the normal bioluminescent reaction, may be part of a still unknown neurochemical mechanism by which the nervous system of the firefly controls the timing of the luminescence.

THE BIOLOGICAL SIGNIFICANCE OF LUMINESCENCE

As we have noted, the value of luminescence in assisting the mating of fireflies is obvious. The same can be said for some marine worms whose reproductive pattern includes a "fire dance."

For fish with organs for luminosity the advantage might be that of intraspecies recognition or of either luring or frightening other animals. Some fish have their own luminescent system, chemically similar to that of *Cypridina,* but others harbor cultures of luminous bacteria in luminous organs. The advantage to the bacteria is that they are supplied with nutrients by the fish. This situation poses a curious problem in microbiology. In laboratory cultures of luminous bacteria there are occasional mutations toward nonluminescence, often associated with luciferase deficiency. Under continuous serial transfer of such cultures, where a little of a fully grown culture is added to a flask of fresh culture medium in order to start a new culture, the "dim" or "dark" mutants tend gradually to displace the "bright" form. This is combated in the laboratory by growing isolated colonies on culture plates, each colony descended from a single cell, and picking a bright colony. But how does the fish maintain a bright culture of bacteria against the mutational drift toward dark forms? Hastings has suggested that the cell wall composition changes with the mutation, so that the fish can recognize and eliminate the dark mutants by specific antigen-antibody reactions.

These considerations expose a more difficult problem centered in the continued existence of luminous bacteria in the seas, not harbored by fish, after countless generations. The same problem can be stated for luminous fungi: Unless there is an ongoing advantage in luminescence, the mutants that do not burden themselves by making the enzyme and substrate for luminescence should have a minute advantage in growth rate and should (in not too many generations) displace the luminous forms. This argument makes it difficult to assert that luminescence is nothing more than an evolutionary relic, left over from a time when the sudden appearance of O_2 in the atmosphere demanded the invention of a mechanism to dispose of O_2. If luminescence had value as a detoxifying device then, it might have that value now. But most bacteria and fungi exist in the presence of O_2, gaining their energy from respiration, without any help from a luminescent system.

Perhaps a biochemical cycle of peroxidation and cleavage, like that in firefly luminescence, is a widespread detoxifying mechanism among animals, bacteria, and fungi, but is usually "dark." In only a few special cases do the requirements of spin and symmetry conservation dictate that a product will be formed in its excited state. In that case the luminescence is an occasional accident of the fact that this kind of chemistry involves a high concentration of energy in a molecule. This "chemical accident" has then been exploited by some animals for other biological purposes.

The chemistry of luminescence might also involve singlet excited O_2 as an intermediate. Hastings has emphasized that the high concentration of energy in a bioluminescent system, especially if singlet O_2^* is formed, might provide a wedge in a metabolic attack on nutrients (sources of carbon for growth) that are relatively resistant toward chemical alteration.

In conclusion, we have in bioluminescence a phenomenon in which the chemistry is becoming clear but the biology remains, for many organisms, quite mysterious.

BIBLIOGRAPHY

1. Harvey, E. N.: "Bioluminescence," Academic Press, Inc., New York, 1952. Although a little out of date, this is a classic ac-

count of bioluminescence by the foremost investigator of the subject during the first half of the century.

2. Seliger, H. H., and W. D. McElroy: "Light: Physical and Biological Action," Academic Press, Inc., New York, 1965. The chapter on bioluminescence contains many interesting details.
3. Hastings, J. W.: Bioluminescence, *Ann. Rev. Biochem.,* vol. 37, p. 597, 1968. A professional review of bioluminescent systems emphasizing the bacteria.
4. McElroy, W. D., H. H. Seliger, and E. H. White: Mechanism of Bioluminescence, Chemiluminescence and Enzyme Function in the Oxidation of Firefly Luciferin, *Photochem. Photobiol.,* vol. 10, p. 153, 1969.

Index

Page references followed by **F** indicates Figure; followed by **S** indicates structural formula.

Absorption spectrometry, 42
Accessory pigments in photosynthesis, 34, 76
Acetabularia, 170
Actinomycin D, 169
Action potential, 127
Action spectrum:
  for chlorophyll fluorescence, 37
  for circadian clock setting, 167
  for the fast photovoltage, 120
  for photoreactivation, 197
  for photoregulation of seed germination, 158
  for photosynthesis, 37, 43, 75
  for phototaxis, 71, 73, 81
  for phototropism, 81
  for phytochrome conversion, 158
  for sunburn and tanning, 202, 203
  for ultraviolet damage, 178
  for vision, 109**F**
Adenine, 181**F,** 181**S**
Adenosine triphosphate (ATP), 11**F,** 11**S**
  and luciferin activation, 224
  and muscle contraction, 83
  reaction with dehydroluciferin, 228
Aerotaxis, 71
Airth, R. L., 218
Alexander the Great, 155
Allard, H. A., 152
Allen, F. L., 49
Allen, M. B., 42
Amacrine cells, 90, 129, 134
Anderson, J. M., 42
Androsthenes, 155
Arnold, W., 27, 30, 50
Arnon, D. I., 16
Atmosphere, primeval earth, 189
Auxins, 163
  role in phototropism, 80

Bacteriochlorophyll, 13**F,** 13**S,** 22, 32, 75
Barlow, H. B., 134
Baumgardt, E., 101
Bayes, K. D., 211
Benoit, J., 149
Benson, A. A., 17
Berends, W., 180
Berkner, L. V., 189
Beukers, R., 180
Binomial statistical distribution, 191
Biological clock (*see* Circadian rhythm)

Bioluminescence (*see* Luminescence)
Bipolar cells, 90, 129, 134, 135
Bishop, N. I., 41
Blind spot, 90**F**
Blinks, L. R., 35, 37
Blue-green algae, 37, 70
Boardman, N. G., 42
Borthwick, H. A., 152
Bouman, M. A., 101
Brown, K. T., 118
Browne, R. J., 207
Bruce, V. G., 167
Buder, J., 71
Bünning, E., 144
Butler, W. L., 49, 157

Calvin, M., 17
Calvin-Benson cycle, 17
Carotene, 74**F, **74**S**
Carotenoid pigments, 33
  absorption band shift, 63
  protective action, 200
Carrier, W. L., 195
Castle, E. S., 78
Chance, B., 42, 54
Chlamydomonas, 41
Chlorella, 27, 30, 50
Chlorobium chlorophyll, 32
Chlorophylls *a* and *b,* 13**F,** 13**S,** 34, 43
Chloroplasts, 55
Cholodny, N., 80
Chromatic transients, 36
Chromatophores (bacterial), 22, 25**F**
Chromophore, 93
Circa-annual rhythm, 142
Circadian rhythm, 140
  in animals, 146
  in *Gonyaulax,* 165, 166**F**
  in microbes, 155
  in *Neurospora,* 141
  in plants, 155
Color blindness, 125
Color discrimination, 123
Color vision, 122
Cone, R. A., 119
Cones (visual), 93, 96
  and color vision, 96
  pigments (human), 123
Contrast and movement detectors, 134, 135
Corey, E. J., 211
Cormier, M. J., 218, 221
Corneal transplant, 137
Crick, F. H. C., 180
Crofts, A. R., 62
Cytochromes, 48
  in *Chromatium,* 52
Cytosine, 181**F,** 181**S**
  photochemical hydration, 179

Dark adaptation, 194, 105**F**
Dehydrocholesterol, 176
Delayed fluorescence of chlorophyll, 50
  and membrane potential, 62
Delbrück, M., 78

Deoxyribonucleic acid (DNA), 169, 178, 180**F**, 180**S**
  dark repair enzyme, 185, 195
  (*See also* Repair of DNA)
Dermis, 202
de Vries, H., 101
Dichlorophenyldimethylurea (DCMU), 41, 48
Dormins, 163
Dowling, J. E., 92, 129, 134
DuBois, R., 216
Dulbecco, R., 196
Dure, L. S., 221
Duysens, L. N. M., 22, 37, 40, 42

Early receptor potential (*see* Fast photovoltage)
Ebrey, T. G., 119
Ehret, C. H., 169
Electron spin, 206
  conservation in chemical reactions, 211
Electroretinogram, 118
Emerson, R., 27, 30, 37
Energy transfer, 37, 64
Engelmann, T. W., 71, 75
Enhancement effect, 37
Epidermis, 202
Ergosterol, 176
Euglena, 71, 73, 167
Eukaryotic organisms, 140
Excited oxygen, 206, 230
  absorption and emission spectra, 207
Eye:
  chromatic aberration in, 99
  horseshoe crab, 128, 132
  human, 89
  squid, 94

Fast photovoltage, 118
Findlay, G. M., 176
Finsen, N., 176
Flavine mononucleotide (FMN), 219, 220**F**, 220**S**
Fleischman, D. E., 62
Flowering, 152
Fluorescence:
  of bacteriochlorophyll, 52, 53
  of chlorophyll in photosystem II, 47
  of decarboxyketoluciferin, 227
  of FMN, 221
  of oxyluciferin, 217
  of P870, 52
Foote, C. S., 211
Fovea, 89, 95
Franck, J., 49
French, C. S., 37
Frenkel, A. W., 16

Gaffron, H., 9, 10, 30
Ganglion cells, 90, 130, 134
Garner, W. W., 152
Gates, F. L., 178
Generator potential, 128
Gest, H., 22

Gibberellins, 163
Gibson, Q. H., 219
Giese, A. C., 178, 200
Glyceraldehyde phosphate, 18**S**
Goldstein, E. B., 119
Gonyaulax, 166, 218
Goodgal, S. H., 197
Gramicidin, 62
Grana in chloroplasts, 55
Granit, R., 128, 132
Grunion, 141
Guanine, 181**F,** 181**S**

Hagins, W. A., 101, 128
Halldal, P., 82
Hanawalt, P. C., 195
Hartline, H. K., 128, 132
Harvey, E. N., 221
Hastings, J. W., 165, 169, 218, 230
Hastings, R., 179
Haxo, F., 37
Haynes, R. H., 195, 200
Hecht, S., 100
Helmholtz, L. von, 122
Hemophilus influenzae, 185
Hendricks, S. B., 152
Herriott, R. M., 197
High-energy reactions in plant photoregulation, 155
Hill, R., 9, 40
Hill reaction, 9
Hillman, W. S., 144, 170
Hoffman-Berling, H., 83
Hollaender, A., 178, 195
Hopkins, T. A., 223
Horizontal cells, 90, 129, 135
Hormones, role in plant photoregulation, 163
Hubbard, R., 113
Hubel, D. H., 134
Hydrogen-adapted algae, 9, 51
Hydrogen bond, 179
Hydrogen transfer, 4, 161
Hydrogenase, 9
Hydroxybutyric acid, 10
Hypophysis (*see* Pituitary gland)
Hypothalamus, 149

Indophenol, 9
Inhibition in visual nerve cells, 107, 132, 135
Iris of eye, 89
  and dark adaptation, 107

Johnson, F. H., 222
Joliot, P., 49

Kaneko, A., 126, 134
Kasha, M., 207
Kearns, D. R., 209, 212
Kelner, A., 196
Keratinocytes, 202, 203
Khan, A. U., 209, 212
Kinetins, 163
Kishi, Y., 217
Kok, B., 26, 42

Kuffler, S. W., 132

Land, E., 126
Lettvin, J. W., 134
Levine, R. P., 41
Lewis, C. M., 37
Light-harvesting bacteriochlorophyll, 22–24
Light-harvesting pigments in photosynthesis, 22, 28, 31, 35**F**
Light-sensitized damage (*see* Photochemical oxidation; Visible light effects)
Links, J., 72
Lipmann, F., 10
Luciferase, 216
Luciferin, 216
  of *Cypridina,* 217**F,** 217**S**
  of firefly, 223**F,** 223**S**
    activation, 223
  of *Renilla,* activation, 218
Luminescence:
  of acorn worms, 221
  of bacteria, 219
    role of aldehyde, 220
  biological significance, 229
  of chlorophyll, 50
  of *Cypridina,* 216
  of fireflies, 222
    assay of ATP, 50
    chemistry, 226**F**
    emission spectra, 222, 228
  of fish, 229
  of fungi, 218
Luminescence:
  of *Gonyaulax,* 218
  of jellyfish, 222
    role of calcium, 222
  of *Renilla,* 218
Lumirhodopsin, 115
Lupus vulgaris, 176

McCapra, F., 223
McElroy, W. D., 223, 228
MacNichol, J, D., 123
Mallet, L., 206
Malpighian layer, 202
Manten, A., 71
Marshall, L. C., 189
Matthews, R., 113
Maturana, H., 134
Maxwell, J. C., 122
Melanin, 203
  formation in prompt tanning, 203
Melanocytes, 202, 203
Membrane:
  electric potential, 57
  ion pumping, 57
  photosynthetic, 55, 56**F**
Meyerhof, O., 10
Microelectrodes, use in nerve studies, 128
Migration, 146, 147
  transequatorial, 148
Mimosa, 154, 162
Molisch, H., 71
Morita, S., 53
Movement and contrast detectors, 134, 135

Mud puppy, 92, 129
Muhammed, A., 197
Murakami, M., 118

Nerve impulse (*see* Action potential)
Nicotinamide adenine dinucleotide (NAD), 12**F,** 12**S,** 21, 108, 218, 219, 222
Nicotinamide adenine dinucleotide phosphate (NADP), 12**F,** 12**S,** 21, 42
Nigericin, 62

Ogryzlo, E. A., 207
Opsin, 93, 111
Optic nerve, 89
  on, off, and on-off responses, 132
Østerberg, G., 90
Oxidation, 2
Oxygen, excited (*see* Excited oxygen)
Oxygen poisoning, 189
Ozone, absorption of ultraviolet by, 33, 189

P700, 26, 41, 48, 51
P800, 24
P840, 32
P870, 24
  oxidation of, 24, 51
Pak, W. L., 119
Pararhodopsin, 115
Parity (*see* Symmetry of wave functions)
Parker, M. W., 152
Parson, W. W., 54
Peroxidase, 221, 230
Phosphatidyl ethanolamine, 112
Phosphoglyceric acid, 18**S**
Phosphorylation:
  coupling to electron transport, 57
    chemical hypothesis, 57
    chemiosmotic hypothesis, 58
  cyclic and noncyclic, 20, 43
  oxidative, 15
  photosynthetic, 16
  uncouplers, 61
Photochemical electron acceptor, 26, 46
Photochemical electron transfer, 3**F,** 26, 203
  time needed, 54
Photochemical ion pumping, 59
Photochemical isomerization, 111, 161
Photochemical oxidation, 161, 204
  of chlorophyll, 200, 204
    protection by carotenoid pigments, 200
  role of triplet state, 204
Photochemical oxygenation, 205
  role of excited oxygen, 206
Photochemical reaction centers, 22, 25**F**

Photoperiodism, 142
  and the circadian clock, 144, 170
  hourglass model, 143, 165
Photoprotection, 200
Photoreactivating enzyme, 197
  chromophore, 198
Photoreactivation, 170, 186
  distribution in nature, 196
  and pyrimidine dimers, 197
  of transforming DNA, 196
Photoreceptor:
  for photoregulation: in animals, 149
    in insects, 169
    in plants (phytochrome), 150
  for phototaxis, 71
  for phototropism, 80
  for vision in worms, 86
Photoregulation (*see* Circadian rhythm; Photoperiodism; Seasonal regulation)
Photostationary state, 115
  of phytochrome, 150, 153
  in thymine dimerization, 183
  of visual pigment, 117
Photosynthesis:
  bacterial, 6, 10, 51
  carbon assimilation, 17
  overall reaction, 5
  as oxidation reduction, 7
  quantum efficiency, 30
  redox potentials, 21, 52
Photosynthesis:
  two photosystems in, 19, 20, 34, 39, 44**F**
    in bacteria, 53
Photosynthetic bacteria, 6
  cell membrane, 54
Photosynthetic oxygen evolution, 19, 43
  activation effect, 49
  yield of flash, 27, 30, 49
Photosynthetic reaction center (*see* Photochemical reaction centers)
Photosynthetic unit, 29
Photosystem I, 19, 26, 51
Photosystem II, 19
  electron donation to, 49
Phototaxis:
  of chloroplasts, 68
  hynothesis of Links, 72, 82
  and ions, 82
  of microbes, 68, 167
  phobic and topic responses, 70
Phototropism:
  of oat seedlings, 79
  of *Phycomyces,* 76
Phycobilins, 34, 37
Phycocyanin, 34, 157
Phycoerythrin, 34
Phycomyces:
  light growth response, 78
  structure, 76
Phytochrome, 150, 159**F**, 159**S**
  absorption spectrum, 158
  chemistry, 156

Phytochrome:
photochemistry, 160, 162**F**
Pigment epithelium in eye, 89
Pineal organ, 148
Pirenne, M. H., 100
Pittendrigh, C. S., 144
Pituitary gland, 149
Plant hormones (*see* Hormones)
Plastocyanin, 42, 48
Plastoquinone, 47
Poisson statistical formula, 102
analysis of visual excitation, 102
target theory for radiation effects, 191
Polyak, S. L., 90
Porphyria, 176
Prelumirhodopsin, 114
Purines, 181**F,** 181**S**
Pyrimidines, 181**F,** 181**S**
Pyrophosphoric acid, 224**S**

Quantum efficiency:
of amino acid reactions, 188
for photochemical oxygenations, 205
of photosynthesis, 30
for phytochrome conversion, 158, 159
for thymine dimerization, 183
of visual excitation, 101

Raab, O., 176
Radiation effect versus dose, curves, 194**F**
Ramon y Cajal, S., 90
Receptive field, 134
and dark adaptaiton, 107
Receptor cells in vision, 89, 90, 129, 135
depolarization and hyperpolarization, 129
sodium ion current, 129
Recovery from radiation damage, 195
Red algae, 37
Red drop effect, 37
Reduction, 2
Reductive dephosphorylation, 16, 63
Repair of DNA, 185
in bacteria, 187, 195
in skin, 195
and thymine dimers, 195
Respiration, 7, 14
Retina, 86, 89
of squid, 94, 128
Retinal, 93, 109
isomers, 111, 112**F**
Retinene (*see* Retinal)
Retinol (*see* Vitamin A)
Rhodopseudomonas, 24, 32, 54
Rhodopsin, 93, 108
absorption spectrum, 109**F**
bleaching and dark adaptation, 106
chemistry, relation to vitamin A, 113
Rhodospirillum, 53
phototaxis of, 70

Riboflavin, 74**F**, 74**S**
Ribonucleic acid (RNA), 169, 178
Ribulose diphosphate, 18**S**
Rods (visual), 93, 96
Rose, A., 89
Rupert, C. S., 185, 197
Rushton, W. A. H., 106, 123

Saiga, Y., 222
Schiff base linkage, 112
Schlaer, S., 100
Schmid, G., 30
Scintillons, 218
Sea hare (*Aplysia*), 170
Seasonal regulation, 141
in animals, 147
(*See also* Photoperiodism) 141
Seliger, H. H., 207, 223, 228
Sensitive plant (see *Mimosa*)
Series formulation (*see* Photosynthesis, two photosystems in)
Setlow, J. K., 185, 195, 197
Setlow, R. B., 185, 195, 197
Shimomura, O., 222
Siegelman, H. W., 157
Sieve effect, 203
Sinsheimer, R. L., 179
Sistrom, W. R., 26
Skeleton photoperiod, 144
Skin, human: anatomy, 202
responses to light, 202, 203
Smith, K. C., 188
Stanier, R. Y., 22
Stiles, W. S., 123
Stratum corneum, 202
Strehler, B. L., 50
Subjective day and night, 144
Sunburn, 202
Sweeney, B. M., 166
Sybesma, C., 53
Symmetry of wave functions, conservation in chemical reactions, 212
Synapse, 92, 127

Tanning, 203
Target theory in radiation biology, 190
Threshold for vision, 100, 104
Thylakoids:
of chloroplasts, 55
of photosynthetic bacteria, 55
of visual receptor cells, 93
Thymine, 181**F**, 181**S**
photochemical dimerization, 180
Tomita, T., 129, 134
Totter, J. R., 218
Transforming DNA, 185

Ultraviolet effects:
on amino acids, 188
bactericidal action, 176, 178

Ultraviolet effects:
- dimerization of thymine, 183
- enzyme inactivation, 187
- skin cancer, 176
- sunburn and tanning, 202
- thickening of skin, 203
- transforming DNA inactivation, 186
- vitamin D formation, 176

Uracil, 181**F**, 181**S**

Valinomycin, 62
van der Velden, H. A., 101
van Niel, C. B., 6–9
Visible light effects:
- drug sensitization, 176, 201
- dye sensitization, 176
- role of oxygen, 200
- role of triplet excited states, 199
- (*See also* Photochemical oxidation)

Visual acuity, 97
Visual fatigue, 131
Visual information processing, 131
Visual pigment, 93, 108, 111
- photochemical cycle, 113, 116**F**, 161
  - kinetics in rat eye, 122
  - in squid eye, 118

Visual pigment:
- polarization, 94
- (*See also* Rhodopsin)

Visual receptors (*see* Cones; Receptor cells in vision; Rods)
Visual sensitivity, 96, 100, 106
Visual transduction and sodium ion current, 130
Vitamin A, 108
Vitamin D, 176
von Helmholtz, L., 122

Wang, S. Y., 179
Watson, J. D., 180
Went, F., 80
White, E. H., 223, 228
Wiesel, T. N., 134
Wilkins, M. H. F., 180
Witt, H. T., 26, 42

Xeroderma pigmentosum, 195

Yamashita, T., 49
Yoshizawa, T., 113
Young, T., 122

Z-scheme (*see* Photosynthesis, two photosystems in)